Roentgen Diagnosis of Children

A Self-Teaching Manual

M. EIKEN

Director, Department of Radiology,
Gentofte Hospital,
University of Copenhagen.

Consultant Radiologist,
The Children's Hospital, Fuglebakken,
Copenhagen.

An F. A. D. L. s FORLAG A. S. publication

Distributed by
Year Book Medical Publishers, Inc.
Chicago · London

Original Danish Edition
Copyright 1977 by F.A.D.L.s Forlag
English Edition
Copyright 1977 by F.A.D.L.s Forlag
Translated from the Danish by Harry Cowan, B.Sc.
This book is copyrighted in Denmark and may not be
reproduced by any means in whole or in part without written
permission from the copyright owner.
Distributed throughout the world except for Denmark, Norway,
Finland and Sweden by
Year Book Medical Publishers, Inc.
Library of Congress Catalog Card Number: 77-81871
ISBN: 0-8151-3028-7
by arrangement with
F.A.D.L.s Forlag A.S.
ISBN: 87-7437-630-6
Printed in Denmark
by Th. Laursens Bogtrykkeri A-S

Books in the series:
Roentgen Diagnosis of the Chest
Roentgen Diagnosis of Bones
Roentgen Diagnosis of Children

To

Peter, Birgitte and Henrik

Preface

ROENTGEN EXAMINATION of children primarily is characterized by the many congenital organ changes that, in the majority of cases, can be demonstrated in the infant or already in the newborn. Many acquired diseases are also characteristic of childhood, and not rarely are found associated with definite age groups. The older the children become the more will the roentgen findings resemble what we are familiar with in adult patients. In this book, the main emphasis is on radiologic findings in early childhood. Some of the findings demonstrated are quite uncommon but have been included to throw light on problems of differential diagnosis. For reasons of space, neuroradiologic problems have not been included, just as skeletal changes have been touched on only peripherally.

The structure of the book otherwise follows the principles of the earlier volumes in the series, so that, for example, the supplementary radiograms always can easily be compared with the primary views by folding the pages.

M. EIKEN

Contents

Introduction

Thorax

RESPIRATORY ORGANS

A disturbance of respiration is the most common indication for radiologic examination of the chest. During the neonatal period, it is particularly common in the form of respiratory distress.

Hyaline membranes (idiopathic respiratory distress syndrome – RDS) present with increased respiratory rate, retraction of the sternum and the lower part of the thorax and cyanosis. The signs appear immediately after birth, almost exclusively in premature infants and infants of diabetic mothers.

The earliest radiologic findings – which often may be recognized a few hours after birth – are small, closely packed densities (atelectatic alveoli), giving the lungs a characteristically firm, finely granulated structure. The air-filled bronchial branches stand out clearly against a background of numerous alveolar atelectases *(air bronchogram)*. It also is characteristic that the lungs are small and stiff. The structure of the lungs gradually becomes more coarsely meshed with coalescenses of the alveolar atelectases and increasing hyperinflation of the intervening groups of alveoli and bronchioles. The mortality is high, but in the children who survive, the signs usually culminate in the course of 2-4 days, often followed by quite rapid improvement.

The aspiration syndrome, like hyaline membranes, is characterized by varying degrees of respiratory distress, developing immediately after birth. The most serious cases are seen in postmature infants who have aspirated amniotic fluid containing meconium, usually following fetal asphyxia. The radiologic findings may vary from coarse linear densities radiating out from the hilar areas to perihilar patchy densities or to larger or smaller wedge-shaped atelectases in one or both lungs. The changes are nonspecific and cannot be distinguished from bronchopneumonia or pulmonary hemorrhage. The thorax is either normal in size or expanded, with elevated sternum, and localized obstructive emphysema is seen occasionally. Many of these infants die, often within 24 hours of birth. Otherwise, the clinical and radiologic changes disappear in the course of 1-2 weeks.

In premature and mature infants quite comparable radiologic changes may be found without any sign of meconium having been aspirated. This often is described as *"wet lungs"*. Some of these infants clinically are asymptomatic, some have tachypnea and others, again, have respiratory distress, usually of only a few days' duration. A characteristic feature is the usually rapid regression of the radiologic changes. The pathogenesis is not certain, but aspiration in utero of varying amounts of amniotic fluid containing formed elements such as squamous cells and lanugo hairs probably is a frequent cause of the condition. Other possibilities are incomplete expression of the fluid contents of the lungs during delivery, failing resorption of the remaining pulmonary fluid via the lymphatics and capillaries immediately after birth and reduced respiratory activity as a result of oversedation of the mother. The radiologic appearance of a complicating pneumonia, which may occur in the course of both RDS and the aspiration syndrome, may be difficult to differentiate from the lung changes already present. *Pneumomediastinum and pneumothorax* are frequent complications and may arise spontaneously or be caused by too vigorous ventilation.

More uncommon changes, such as congenital *lymphangiectasia, pulmonary hemorrhage, abnormal pulmonary venous return* and occasionally *congenital heart disease* with left-to-right shunt, may result in uncharacteristic radiologic findings, which may be impossible to distinguish from those mentioned above.

Respiratory distress may also be observed in the presence of elevated diaphragmatic domes due to space-occupying lesions in the abdomen or in pneumoperitoneum, in thoracic deformities (Jeune's disease, osteogenesis imperfecta Vrolik and other conditions) and in intra-thoracic space-occupying lesions. In cerebral asphyxia or in cerebral hemorrhage, respiratory disturbances often are characterized by periods of apnea.

Wilson-Mikity's syndrome (pulmonary dysmaturity) usually does not cause any signs until a few days to weeks after birth, when the infants become dyspneic, with attacks of cyanosis. This syndrome occurs primarily in prematures. Diffusely scattered in both lungs there is a coarsely meshed, reticular pattern with emphysema of the intervening lung tissue. The lungs appear to be of normal size or overexpanded. The clinical and radiologic changes may disappear gradually in the course of 3–12 months, the radiologic changes usually somewhat later than the clinical signs.

Pneumonia occurs with particular frequency in infants under the age of 1 year. In newborns and infants, two main types can be distinguished clinically and radiologically, an *alveolar* or exudative form, characterized by plentiful amounts of inflammatory exudate in the alveoli, and an *interstitial* or proliferative form, where the inflammatory changes mainly are localized in the bronchial walls, with proliferation of the peribronchial and interstitial tissue, but only limited amounts of exudate.

A combination of the two types, a *mixed pneumonia,* is rather common.

Alveolar pneumonia, which may occur as bronchopneumonia, segmental pneumonia or more rarely lobar pneumonia, mainly is caused by bacteria, usually pneumococci, streptococci, staphylococci, *Hemophilus influenzae* and Klebsiella (Friedländer). Accordingly, the radiologic findings may vary from multiple smaller or larger, round, ill-defined and often confluent consolidations in one or both lungs, nonhomogeneous segmental consolidations, sometimes sharply delimited by one of the interlobar fissures, or lobar, often intense, consolidations.

Interstitial pneumonia is caused predominantly by viruses but also plasma cell pneumonia and fungal infections mainly are of this type. Clinically, they often are classified as primary atypical pneumonia. The radiologic findings are characterized by the increased bronchovascular markings. The changes may be localized mainly to the perihilar regions or may be distributed diffusely throughout both lungs, of very varying intensity. During the course of the disease, the inflammatory exudate in the bronchial branches may cause local consolidation of the lung tissue or obstructive emphysema. The radiologic picture thus may change rapidly, and it is most easy to make a differential diagnosis in an early stage of the disease.

Hilar lymphadenitis may accompany both the alveolar and the interstitial pneumonias, but as the hilar areas often are enlarged as a part of the parenchymal changes, the lymph nodes may not be seen.

Pneumonia may occur even in the newborn following intrauterine aspiration of infected amniotic fluid, and then it presents the same radiologic findings as in the aspiration syndrome. *Empyema*, solitary or multiple *pulmonary abscesses* or the development of one or more cystic, air-filled cavities in the lung tissue – *pneumatoceles* – particularly are observed in infections with staphylococci. Pneumatoceles may become very large, but almost always disappear spontaneously.

Tuberculosis. The classic radiologic picture of primary tuberculosis of the lungs is a small consolidation in one of the lungs, a linear lymphangitic density directed in toward enlarged, regional lymph nodes in the hilum and a local pleural thickening. However, such a primary complex seldom is observed, as the changes rapidly are hidden by a lobular or lobar increase in density of perifocal alveolar exudate, which cannot be distinguished radiologically from consolidations of other bacterial origin. Pronounced enlargement of hilar lymph nodes increases the suspicion of tuberculosis. Fungal infections, especially moniliasis, coccidioidomycosis and histoplasmosis, may give rise to pulmonary findings that can easily be mistaken for primary tuberculosis.

Bronchitis. In asthmatic bronchitis, spasm in the peripheral bronchi will result in the development of obstructive emphysema. A pronounced increase in mucus secretion may cause bronchial obstruction, and regional atelectases therefore are seen often. In acute cases, the emphysema is reversible. In chronic bronchitis, the hilar areas become more prominent because of the increased bronchovascular markings. The bronchial branches are dilated, with development of cylindrical or saccular bronchiectases. Both acute and chronic bronchitic changes usually are more striking in the younger age groups. In acute exacerbation – status asthmaticus – the lungs may show extreme expansion, with pronounced depression of the diaphragmatic domes, widening of the intercostal spaces and greatly reduced lung structure peripherally.

Mucoviscidosis (cystic fibrosis of the pancreas) is a generalized disease that, above all, affects the lungs and digestive tract; 10-15 % of the children with the disease have had meconium ileus as newborns. The early radiologic findings in the lungs are nonspecific, characterized by emphysema, enlarged hili and recurrent, bilateral bronchopneumonia. Gradually, as the disease progresses, increasing chronic lungs changes are seen, with line shadows and patchy consolidations due to bronchiectasis, small abscesses and interstitial fibrosis. Localized atelectasis, emphysema and bronchopneumonial consolidations give the lungs a varying appearance. In the late stages, cor pulmonale often develops.

Aspiration of hydrocarbons. Poisoning as a result of aspiration of various hydrocarbons, in particular turpentine and kerosene, particularly occurs in the age group 1-3 years. A number of these children develop patchy, confluent consolidations, mainly in the basal areas of the lungs. The consolidations are nonspecific and cannot be distinguished from alveolar pneumonia.

Foreign bodies in the airways may be retained in the trachea or in a bronchial branch, depending on the size of the body. Nonopaque foreign bodies in the bronchial tree may be demonstrated indirectly, as they often produce a check-valvular obstruction with the development of lobar or lobular emphysema and displacement of the mediastinum. Fluoroscopy shows the mediastinum displaced away from the emphysematous lungs during expiration. In total obstruction, atelectasis develops. The differential diagnosis is made by comparing anamnesis with radiologic findings, possibly supplemented by bronchography and/or bronchoscopy.

Congenital anomalies of the lungs play only a moderate role in the etiology of respiratory difficulties in the newborn. *Agenesis* of one lung, where the lung and its main bronchi are completely missing, or *aplasia*, where rudiments of the bronchial branches are present, may occur without giving symptoms. A radiogram shows the heart and mediastinum lying against one of the thoracic walls with compensatory overexpansion of the remaining lung.

Hypoplasia, with failing development of one or more pulmonary segments, or of an entire pulmonary lobe, is revealed by compensatory emphysema with increased translucency of the remaining pulmonary tissue on that side. When the anomaly is right-sided, it sometimes is accompanied by partial anomalous pulmonary venous return, the abnormal vein presenting as a characteristically arch-shaped vascular shadow, running down along the right margin of the heart to the inferior caval vein – the *scimitar syndrome*.

Congenital lobar emphysema often causes respiratory difficulty in the newborn. The radiogram shows pronounced expansion of part of one lung, most often segments of the left upper lobe, with displacement of the heart and mediastinum to the right and compression of the rest of the left lung. The findings express a check-valvular obstruction of bronchial branches as a result of developmental defects, but in most cases the nature of the defect is not demonstrable.

Pulmonary tumors are relatively rare in children. Among benign tumors is the hamartoma or *cystic adenomatoid malformation,* seen in newborns and infants as non-homogeneous densities or more often as multiple cystic translucencies, usually limited to a single lobe. Among other benign tumors in the chest might be mentioned *teratoma,* usually arising from the anterior mediastinum, *thymoma,* likewise from the anterior mediastinum, and *neurinoma,* usually localized paravertebrally. All these tumors are well delimited and often produce large densities.

Lung cysts almost always are acquired. As a rule, they communicate with the bronchial tree and are seen as air-filled cavities that, by means of a check-valvular bronchial obstruction, may grow rapidly and develop into large tension cysts. In the differential diagnosis, congenital lobar emphysema and cystic adenomatoid malformation must be taken into consideration. *Malignant tumors* almost exclusively are metastases, mainly from nephroblastoma and Ewing's sarcoma.

Physiologic variations and other causes of erroneous diagnosis. The size and shape of the thymus vary considerably in small children. The normal thymus may be very large, increasing the width of the mediastinal shadow and not infrequently blending with and covering large parts of the heart. The size varies with the respiratory cycle, widening during expiration. At times, its contour is wavy, corresponding to the intercostal spaces. Not infrequently a triangular, sail-shaped thymus may be seen to the right of the heart.

Radiograms from the newborn in the supine position often show one or more densities, usually in the lower part of one or both lungs, with a curvilinear lateral border. It may arouse a suspicion of pneumothorax, but is in fact due to skin folds.

Depending on the age of the child, minor diagnostic problems may be caused by the shadows of openings in the incubator cover, of the umbilical cord, a lost comforter or of hair.

The appearance of the lungs also depends largely on the phase of respiration, the lung structure often being considerably denser on films taken in expiration than in inspiration. As far as possible, therefore, routine films of the thorax should be taken in maximal inspiration. The phase of respiration also plays a role in evaluating the size and shape of the heart, just as the phase of contraction of the heart is of significance.

Radiologic examination of the upper respiratory tract is performed to demonstrate suspected changes in the mucosa of the sinuses, especially in children with asthmatic bronchitis. Much more uncommon indications are unilateral or bilateral *choanal atresia,* a membranous or osseous occlusion of the posterior part of the nasal cavity and hypoplasia of the mandible *(micrognathia* in the *Pierre Robin* syndrome), where the tongue falls back and obstructs the air passage.

Stenosis of the trachea is found in various forms of vascular ring, in congenital goiter, cystic hygroma of the neck and in a few other rare forms of space-occupying lesions. *Congenital inspiratory stridor* often is due to insufficient development of the cartilage of the epiglottis, and at times may result in respiratory difficulty, especially in connection with infections of the respiratory tract.

A tracheal bronchus originating from the trachea above the bifurcation, and usually supplying the right apical segment of the upper lobe or ending blind, can be identified only by bronchography.

HEART

Congenital malformations of the heart in newborns are found in a large number of variations. About half of them are diagnosed during the first few weeks of life, and most of them within the first year. However, between 25 % and 33 % of the infants – mainly those with complex malformations – are lost within the first year of life, so that the number of possible types is thereafter reduced.

An early diagnosis is based on the clinical examination, supplemented by plain films of the chest, and with the aim of establishing the nature and degree of severity of the hemodynamic disturbances. If it is determined that operative treatment is indicated, cardiac catheterization and angiocardiography are performed for the definite elucidation of the morphologic changes.

Early radiologic identification comprises an evaluation of the relative size of the heart or, better still, an estimation of the volume of the heart. Further investigations include an analysis of the shape of the heart and the large vessels, demonstration of any possible rotation or displacement of the heart and an evaluation of the pulmonary vascularity. It should be added that the smaller the infant is the more difficult it is to demonstrate any possible radiologic changes.

Only a brief review will be given here to the characteristic findings in some of the more usual forms of congenital heart disease, in those cases where the hemodynamic changes are severe enough to have any effect at all on the appearance of the heart. Together, these malformations constitute more than 80 % of all heart malformations.

Ventricular septal defect (VSD) is the most common cardiac malformation. The heart becomes enlarged with expansion of both ventricles and the left atrium. The main pulmonary artery dilates together with the central sections of the pulmonary arteries, but with a rapidly decreasing caliber toward the periphery.

Atrial septal defect (ASD) also belongs among the most frequent anomalies, often combined with other malformations. Moderate enlargement of the heart, with expansion of the right atrium and ventricle, is found in both ostium primum and ostium secundum defect. The main pulmonary artery and its branches are dilated and the central pulmonary vascularity increased.

Patent ductus arteriosus (PDA) results in enlargement of the left atrium and ventricle, whereby the cardiac apex rotates slightly posteriorly and is hidden under the contour of the diaphragm. The pulmonary vascularity usually is increased, but later may be scanty due to increasing pulmonary vascular constriction.

Transposition of the great vessels, combined with ASD or PDA, causes a slight enlargement of the heart, which increases rapidly. The shape resembles that of an obliquely placed egg with the pointed end corresponding to the cardiac apex. The base, composed of the great vessels, is quite narrow on the frontal film, broad on the lateral film, with the aorta arising anteriorly. The pulmonary vascularity is increased.

Fallot's tetrad shows accentuated prominence of the hypertrophied and dilated right ventricle, but the size of the heart most often is normal. The segment of the main pulmonary artery is concave (pulmonary window) and the pulmonary vascularity usually is decreased. The apex of the heart often is characteristically elevated and the aortic arch has a right-sided course in 25 % of cases.

Valvular pulmonary stenosis with intact ventricular septum may result in increased convexity of the right heart contour, due to a hypertrophy of the right ventricle, but more decisive is the demonstration of a post-stenotic dilatation of the main pulmonary artery whereas the pulmonary vascularity is normal or reduced.

Valvular and *subvalvular aortic stenosis* give no early radiologic changes. The valvular stenosis gradually causes a dilatation of the ascending aorta, just as the longitudinal axis of the heart is prolonged down to the left as a result of hypertrophy of the left ventricle. The pulmonary vascularity is normal.

Coarctation of the aorta may produce no radiologic changes in the newborn or infants, even though, particularly in this age group, the heart at times may be clearly enlarged. It is not until after the age of 3-4 years that coarctation occasionally may be observed as a constriction of the aorta, immediately below the arcus, possibly followed by a visible poststenotic dilatation. Notching of inferior borders of the ribs usually first becomes visible at the age of 8-10 years. The diagnosis usually is made in the first instance by the weak or missing pulsation in the femoral arteries.

Numerous other developmental anomalies of the aortic arch and the pulmonary artery are known, among which might be mentioned *vascular ring* in the form of double aortic arch, right-sided aorta with left-sided ligamentum arteriosum, abnormal origin of the brachiocephalic trunk or of the right subclavian artery and abnormal origin of the left pulmonary artery from the right. They all may result in compression of the esophagus and trachea and in a number of cases cause laryngeal stridor in infants. It is possible to obtain an idea of the nature of the anomaly from the appearance of the trachea, the esophagus and the localization of the aortic arch. As in the majority of the other congenital malformations of the heart, angiocardiography is decisive for the final elucidation.

Abdomen and Gastrointestinal Tract

Diseases of the gastrointestinal tract are relatively more uncommon in infants than in adults. The congenital malformations dominate among newborns whereas among infants and older children a number of acquired diseases are encountered, more os less characteristic for the different age groups.

ESOPHAGUS

Esophageal atresia is a relatively frequent malformation (1:3000 newborn). Several types are distinguished, depending on the occurrence and localization of communication between the two esophageal segments and the trachea or the main bronchi. In 8 % of atresia there is no fistula to the trachea. In more than 90 % of the cases the upper esophageal segment ends blindly at about the level of the 4th thoracic vertebra whereas the lower segment communicates with the trachea. The dilated upper esophageal pouch, filled with mucus and air, often can be observed on plain films of the chest. The presence of air in the gastrointestinal tract indicates that a fistula to the lower segment must be present.

A fistula between the upper segment and the trachea may be demonstrated by injecting a small amount of water-soluble contrast material into the segment via a catheter.

Aspiration of mucus or food often gives rise to respiratory difficulty, attacks of cyanosis and pneumonia.

In many cases, esophageal atresia is combined with other malformations, especially congenital heart disease or atresia of other sections of the gastrointestinal tract.

Tracheoesophageal fistula without simultaneous esophageal atresia constitutes less than 4 % of all malformations of the esophagus. In the presence of a fistula, the stomach and small intestine often will clearly be distended by air. A frequent reason for clinical suspicion of a fistula is immature swallowing reflexes, especially in premature infants.

Esophageal stenosis of congenital origin must also be regarded as extremely uncommon. It may be localized at the same level as the atresias or involve the distal part of the esophagus. Several of the stenoses diagnosed in infancy are more likely to be secondary to chronic esophagitis, but certain radiologic differentiation is not possible.

Vomiting in the newborn and infants, when starting shortly after birth, may be due to either inadequate closure of the cardiac sphincter (*cardioesophageal relaxation* or *chalasia*) of unknown etiology or hiatal hernia, where part of the ventricle slides up through the esophageal hiatus. Some of these cases, at first sight resembling chalasia, reveal, on closer examination, the presence of a small sliding hernia. In both conditions, reflux from ventricle to esophagus may cause a peptic esophagitis, with possible development of a stenosis of the distal part of the esophagus.

Brachyesophagus probably has developed, in most cases, secondary to hiatal hernia with esophagitis.

In *paraesophageal hernia,* part of the fundus of the stomach slides up into the thorax through the esophageal hiatus whereas the cardiac sphincter remains in the normal position. There is no reflux, no esophagitis and usually no clinical symptoms.

Esophagitis may occur not only from reflux of the gastric contents but as a consequence of swallowing corrosive compounds, most often lye. The radiogram at first shows an irregular contour of the esophagus, possibly with one or more ulcers, in severe cases with the development of secondary stricture, usually of the lower half of the esophagus.

Achalasia or *cardiospasm* is due to inadequate relaxation of the distal end of the esophagus because of defective development of Auerbach's myenteric plexus. It is observed most frequently in slightly older children and is responsible for a greatly delayed passage of contrast material and gradual dilatation of the entire esophagus ending in a downward-pointing conical tip.

Esophageal varices are uncommon but may be seen in portal hypertension with obstruction of the intrahepatic or prehepatic blood flow. They are also seen in tumor of the mediastinum with compression of the azygos vein or the superior caval vein.

Vascular compression of the esophagus with or without difficulty in swallowing is found in developmental anomalies of the aortic arch, the great vessels from the aortic arch and of the pulmonary artery, as mentioned on page 10.

Foreign bodies often are swallowed by small children. Depending on the size and shape of the foreign body, it may be retained in the hypopharynx or in the esophagus. Nonopaque foreign bodies usually may be demonstrated by giving a small amount of contrast material orally. Retention of a foreign body is seen particularly often in congenital or acquired stenosis of the esophagus, including those patients who have been operated on for esophageal atresia.

DIAPHRAGM

Diaphragmatic hernia comprices hiatal hernia (sliding hernia, paraesophageal hernia) and herniation through congenital defects of the musculature of the diaphragm, mainly localized posterolaterally (Bochdalek hernia) or parasternally (Morgagni hernia). Muscular defects in the diaphragm with another localization may occur at times.

Bochdalek hernia most often is demonstrated immediately after birth because the infant has dyspnea. Radiologic examination shows more or less extensive consolidation, usually of the left half of the thorax, caused by the abdominal organs. These may be the small intestine, the spleen, part of the colon and, in a relatively few cases, the stomach. The presence of intestinal gas often gives a vesicular structure. At operation, the left lung is found to be atelectatic and incompletely developed, but once the diaphragmatic defect is closed, the lung often will unfold and become normal in the course of a few days to weeks.

Morgagni hernia usually presents as small, homogeneous, well-delineated protrusions above the anterior medial part of the diaphragm on one or both sides. It usually gives no symptoms and is discovered by chance.

Various *functional disturbances of the diaphragm* may result in abnormal elevation of one or both diaphragmatic domes. Meteorism or pneumoperitoneum, particularly in infants, can produce such powerful elevation of the diaphragmatic domes that the respiration is affected, whereas space-occupying intrathoracic lesions, abnormal distention of a lung segment or a pneumothorax can result in depression of the diaphragm.

STOMACH

Congenital malformations of the stomach are very rare. A *situs inversus* may involve both the thoracic and abdominal organs or one of these regions alone. Occasionally in newborns, *spontaneous perforation* of the stomach is found, with pneumoperitoneum, possibly occurring as a result of developmental defects in the musculature of the stomach wall, but more probably on the basis of thromboembolic septicemia. In questions of differential diagnosis, it is necessary to exclude perforation of the small or large intestine in intestinal obstruction, this lesion having a greater incidence than spontaneous perforation of the stomach.

Infantile hypertrophic pyloric stenosis occurs especially in 3-4-week-old infants, who develop increasing attacks of explosive vomiting that is not bile-stained. The diagnosis usually can be made on the clinical examination alone. In cases of doubt, radiologic examination will reveal a pronounced narrowing and elongation of the pyloric canal, increased ventricular peristalsis and delayed emptying.

Foreign bodies, often nonopaque plastic objects, can be made visible by giving a small amount of barium by mouth. This will make is possible to decide whether the foreign body can be expected to pass spontaneously. Occasionally, larger or smaller filling defects in the barium shadow of the stomach are seen, composed of solid masses of hair and other indigestible components – a *trichobezoar.*

Gastric ulcers are rare in early infancy, somewhat more frequent after the age of 10 years. The ulcers often are quite small, and secondary changes in the form of edema and spasm are not very obvious. *Malignant gastric tumor* is almost never seen in children. On the other hand, benign solitary *polyps* are seen occasionally.

SMALL INTESTINE

Congenital anomalies of the intestinal tract are closely associated with the occurrence of neonatal intestinal obstruction. The distribution of air in the intestinal tract and the degree of distention provide information as to the approximate localization of the obstruction. The necessary background for the demonstration of pathologic changes in the gastrointestinal tract, therefore, is a knowledge of the normal distribution and amount of air in the tract. Normally, air is seen in the stomach immediately after birth. Four hours later, the entire small intestine usually contains air, and after 6-9 hours, the air has reached down to the distal part of the colon. However, in many cases it is exceedingly difficult to distinguish air-filled colonic segment from small intestine.

Atresia and stenosis may involve any portion of the *duodenum, jejunum* or *ileum* and not uncommonly is multiple. The more distal the site of the obstruction the more difficult it is to determine its localization, as the distal, dilated intestinal loops often contain fluid and therefore cannot immediately be recognized as pre-obstructive. In duodenal obstruction there usually is dilatation of the stomach, and in particular of the pre-stenotic part of the duodenum. In other obstructions of the small intestine it is characteristic that the more distally the obstruction is localized the more pronounced usually is the abdominal distention. Atresia of the small intestine most often is localized to the distal part of the ileum. The next most common localization is the duodenum. Besides segmental atresias, duodenal obstruction may appear in the form of a thin diaphragm with or without a central opening, annular pancreas and rotational anomalies.

Obstruction of the small intestine, moreover, can result in a strongly varying picture. In some cases, the passage may be so narrow that the radiologic findings correspond to those seen in atresia. In other cases, a plain film of the abdomen shows no or only doubtful signs of obstruction, and an examination using contrast material is necessary for the demonstration.

Duplications of the alimentary tract result in spherical or tubular cystic structures with smooth musculature in the wall and an internal covering of mucosa of esophageal, gastric or intestinal origin. These structures may be localized in relation to any section of the alimentary tract, independent of their mucosal covering, but are found most commonly in relation to the ileum. They communicate very rarely with the alimentary tract, and therefore are demonstrated only as space-occupying lesions that dislocate and possibly compress the stomach or intestines. An intrathoracic duplication usually causes respiratory difficulties and is observed as a well-defined density, usually to the right of the esophagus and often combined with anomalies of the vertebrae.

Meconium ileus develops when the meconium is too viscid to pass along the intestine, and often is the first sign of cystic fibrosis of the pancreas. In up to half of the cases, meconium ileus is accompanied by ileum volvulus or meconium peritonitis. In the latter case, the intestinal wall has been perforated and meconium has passed into the peritoneal cavity. The presence of calcifications along the lower border of the abdominal cavity suggests perforation. In pneumoperitoneum of the newborn, meconium ileus with perforation must be kept in mind.

Abnorminal hernia occurs both in the newborn and older children and is partly umbilical, partly inguinal. Inguinal hernia occasionally is observed on a plain radiogram of the abdomen, as air-filled intestinal loops may enter the hernial sac. If the hernia causes an obstruction, the radiologic findings are similar to those in other forms of low intestinal obstruction, and a differential diagnosis is possible only if intestinal air can be demonstrated in the hernial sac. Herniography with injection of water-soluble contrast material into the peritoneal cavity permits the demonstration of hernias that are difficult to diagnose clinically.

Duodenal ulcer is more common than gastric ulcer, but nevertheless is demonstrated in only a very few cases in infants, and even in older children it may be difficult to recognize. Just as in gastric ulcer, the craters are small and the secondary deformity slight. Persistent pylorospasm should arouse suspicion of duodenal ulcer.

Interposition of air-filled intestinal loops, especially the colon, between the diaphragm and the liver (Chilaiditi's syndrome) usually is intermittent and due to air swallowing with meteorism.

Necrotizing enterocolitis may involve both small and large intestine and be more or less localized. The etiology is not known, but the lesions are seen particularly in infants with hyaline membranes, so that hypoxia of the intestinal wall has been considered a possible cause. The necrotic lesions of the intestinal wall may cause perforation, and pneumoperitoneum may be the first radiologic sign. In other cases, air is seen in the wall of one or several intestinal segments with an appearance as in pneumatosis intestinalis in adults.

Terminal ileitis is the term for Crohn's disease localized to the distal portion of the ileum. Chronic inflammation with spasms and cicatricial changes result in stenosis of the affected intestinal segment, the contour becomes characteristically serrated and the passage of contrast material delayed. Not rarely, the process extends into the cecum and the ascending colon.

Intussusception of an intestinal segment into another is seen three times as freqently in boys as in girls, and particularly in the age group from 5 to 12 months. Ileoileal or jejunoileal invaginations constitute less than 10 % of the total number of invaginations. The radiologic examination usually shows an obstuction of the small intestine, and the child may have violent attacks of vomiting, but the differential diagnosis most often is first established at operation. The situation is different in the case of the ileocolic and ileoileocolic invaginations, the former of the two representing 75 % of all cases. Meckel's diverticulum, polyps or enlarged lymph nodes may be the cause of the invagination, but usually no cause can be demonstrated. The diagnosis is easily made by a barium enema, showing the characteristic picture of the head of the invagination with complete colonic obstruction. The hydrostatic pressure of the enema often makes it possible to reduce the invagination.

Ascaris lumbricoides is found mainly in slightly older children. If barium is given by mouth, the organisms are seen either as tubular translucencies in the contrast material or as thin barium streaks in the enteric channels of the worm immediately after the contrast material has passed through the patient's small intestine.

Tumors are either benign, solitary adenomatous polyps or multiple polyps in the Peutz-Jeghers syndrome (familial intestinal polyposis), where, in addition, there is characteristic pigmentation of the oral mucous membrane, lips, hands and feet. It can be exceedingly difficult to demonstrate the presence of polyps in the small intestine radiologically, even in the multiple form, as the polyps often are only a few millimeters in diameter.

Malabsorption and *intestinal allergy* (celiac disease, milk allergy, gluten intolerance) at times may result in pronounced radiologic findings. A barium meal passes through the duodenum and upper part of the jejunum relatively rapidly, but the mucous membrane appears to be hypertrophic, with coarsening of the mucosal pattern. The distribution of barium in the jejunum and ileum may show pronounced discontinuity, as short, dilated, contrast-filled segments alternate with segments showing spastic contraction, so-called segmentation, which is the most characteristic finding. The over-all rate of passage through the small intestine often is prolonged.

However, the changes are inconstant. They may be completely lacking or can be so modest that they are difficult to recognize. In other cases, an intestinal infection (acute gastroenteritis) may give quite corresponding radiologic changes. The criteria cannot be used in the newborn and infants, as normal children in this age group often show a pronounced segmentation of barium in the small intestine.

LARGE INTESTINE

Atresia of the colon almost always is localized to the rectum. A distinction is made between high atresia and low atresia, depending on whether the atresia begins above or below a line from the symphysis to the lower edge of the 5th sacral vertebra. In the case of high atresia, associated malformations often are present, primarily in the form of a fistula from the proximal rectal segment to the urethra in boys or to the vagina in girls. Skeletal anomalies are common, especially in the vertebral column. In low atresia there may be a fistula from the rectum to the perineum. In some of the cases of low atresia there is only a thin membranous septum.

Congenital megacolon (Hirschsprung's disease) is due to defective development of the myenteric plexus in the wall of the large intestine. The aganglionic intestinal segment lacks peristalsis and thus inhibits passage of the intestinal contents, resulting in severe constipation with increasing dilatation above the aganglionic segment. The aganglionosis may involve any segment of the colon, but most frequently it is localized to the rectum. Congenital megacolon is considerably more common in boys than in girls. Not infrequently, the obstruction causes a mechanical ileus. The diagnosis may be verified further by anal tonometry and biopsy.

In nonspecific *chronic constipation*, the entire colon becomes dilated and elongated *(functional megacolon)*. At times, these changes may be difficult to distinguish from aganglionosis distally in the rectum.

Anomalies of rotation are due to inadequate rotation early in fetal life of that part of the intestine supplied by the superior mesenteric artery. In typical cases, the duodenum passes down on the right of the vertebral column, the jejunum lies upward in the right side of the abdomen and the ileum downward while the entire colon lies in the left side. This anomaly is called "mesenterium ileocolicum commune" or "nonrotation". At times, the large intestine undergoes secondary rotational movements, whereby the cecum may take up a position upward on the right side or the entire colon can be brought into an apparently normal position. An abnormal course of the duodenum thus is not always tantamount to the appendix lying in the left side of the abdomen.

The abnormal position of the intestine is combined, for one thing, with a very narrow mesenteric-peritoneal attachment, so that volvulus of the small intestine easily develops, and, for another, some abnormal peritoneal strings or bands appear at various sites in the peritoneal cavity (Ladd's bands), the remains of fetal peritoneal membranes. These may cause compression of various intestinal segments, especially the duodenum.

Situs inversus is the term describing a mirror image placement of the abdominal organs. The reversed position usually is noticed by accident and is without significance provided that it is not combined with other significant malformations.

Meconium plug syndrome is due to an accumulation of viscid meconium in the colon, in contrast to the conditions in meconium ileus, where the inspissated meconium lies in the distal part of the ileum. The infants often have passed a small amount of meconium normally, after which a standstill has developed. The abdomen may appear slightly distended and the infant is whimpering. A contrast enema reveals a slightly dilated colon, filled out with meconium masses. The differential diagnosis must take into account in particular the possibility of Hirschsprung's disease, but the further course provides the diagnosis, since so long as it is not a case of Hirschsprung's disease, the enema usually is followed by passage of a plentiful amount of meconium, after which the infant is permanently relieved.

In administering a barium enema to the newborn with a low-placed obstruction of the small intestine, the colon often will appear very narrow. This condition is described as *microcolon*, but it should be noted that it usually is a physiologic condition and that the colon will expand normally when the obstruction is removed. However, a differential diagnostic possibility is aganglionosis of the entire colon.

Inflammatory processes such as *ulcerative colitis* and *Crohn's disease* are rare in early infancy, but occur occasionally in slightly older children. The radiologic findings correspond exactly to those seen in adults. Clinically, *benign lymphoid hyperplasia* may resemble the above diseases, but radiologically it is characterized by multiple, small, uniform polypoid filling defects on the wall of the large bowel, usually umbilicated.

Acute abdomen usually indicates ileus or an ileus-like state. In the newborn, the most frequent causes are atresia, malrotation, megacolon, meconium ileus; in infants, invagination; and in older children, the presence of peritoneal bands. When there is clinical suspicion of obstruction, radiologic examination should be able to clarify whether this is present, its localization and the cause. On films taken in the supine position as well as on films taken with the patient in an upright position, observations are made of the boundaries of the abdomen, the amount and localization of intestinal air, its distribution in the small and large intestine, the degree of distention, the presence of air-fluid levels, increased density because of fluid in dilated intestinal loops or in the peritoneum, abnormal soft tissue shadows, calcifications or foreign bodies, as well as eventually free air in the peritoneum.

A barium enema often may be of differential diagnostic value whereas the administration of contrast material by mouth as a rule should be avoided where there is a suspicion of intestinal obstruction.

Appendicitis is found most frequently in the age group 12-14 years, but about 15 % of the cases occur in children below the age of 5 years, and in these cases often with an especially acute course and early perforation. On radiologic examination, signs of paralytic ileus are observed. Acute gastroenteritis, terminal ileitis and mesenteric lymphadenitis are the most important differential diagnostic possibilities.

Space-occupying abdominal processes in infancy may represent nephroblastoma (Wilms' tumor), neuroblastoma, congenital hydronephrosis, polycystic kidney or intestinal duplication. Splenomegaly and/or hepatomegaly may be seen in infection, in heart disease with failure of the left heart, in reticuloses, as, for example, histiocytosis X, Gaucher's disease, Niemann-Pick disease and in metastases from neuroblastoma, leukemia, malignant lymphoma and Hodgkin's disease. Finally, primary benign tumors of the liver are seen, such as hemangioma, hemangioendothelioma and hamartoma.

Biliary tract and pancreas. Radiologic investigation plays a minor role in demonstrating changes in these organs in infancy. Among congenital anomalies might be mentioned atresia of the biliary tract, resulting in the development of cirrhosis of the liver in the course of a few months. Choledochal cyst arises by dilatation of the ductus choledochus, probably as a result of inadequate relaxation of the sphincter of Oddi. In annular pancreas, the pancreatic tissue surrounds the duodenum and causes an obstruction. Occasionally it is possible to demonstrate calcifications in the pancreas in cystic fibrosis.

Bleeding from the intestinal tract is found in the newborn with necrotizing enterocolitis, in infants with Meckel's diverticulum or invagination, in older children with esophageal varicosities, Crohn's disease, ulcerative colitis, intestinal hemangiomas, benign lymphoid hyperplasia, solitary polyps – mainly in the rectum and sigmoid, Peutz-Jeghers' intestinal polyposis and Gardner's familial polyposis of the colon, the last often combined with multiple osteofibroma and tumors of the cutaneous connective tissue.

Urinary Tract

Disease of the urinary tract is common in childhood, especially in the form of infection. The cause of the disease often is a congenital anomaly, with one or another form of obstruction of urinary flow.

KIDNEYS AND URETERS

Aplasia of a kidney with or without aplasia of the ureter may be difficult to recognize by radiologic investigations. Apparent lack of one kidney is, in most cases, due to nonfunction as a result of obstruction of flow, occlusion of the renal artery or ectopic kidney, and these possibilities must be excluded in the first examination.

Hypoplasia, unilateral or bilateral, is also rare. In older children, too small kidneys more likely are due to degeneration of the parenchyma as a result of acquired renal disease, such as chronic glomerulonephritis (bilateral), chronic pyelonephritis or arterial stenosis.

Fusion of the kidneys occurs particularly in the form of horseshoe kidney with a bridge formation between the lower poles. It is characteristic that the long axes of the kidneys are parallel or converge downward, in contrast to the normal state. The kidneys are inadequately rotated, so that the renal pelvis faces forward. Both on the frontal and lateral views, the ureter has an S-shaped course down over the lower poles of the kidneys.
Fusion of the upper poles or of the midsection of the kidneys (thyroid form) is much less common.

Duplication of a kidney is a very rare occurrence whereas *duplication of the renal pelvis* and of a larger or smaller portion of the ureter – unilaterally or bilaterally – is among the most common anomalies of the urinary tract. When both renal pelves are functioning normally, the diagnosis appears directly from the urography, but should one of the pelves, usually the upper one, not function because of an interruption in the flow, the diagnosis usually can be made by observing that the kidney is longer than normal, that there often is a reduced number of calices in the functioning pelvis and that a line through the peripheral ends of the upper and lower minor calix of the functioning lower pelvis converges with the median plane downward. *Duplication of the ureter* may comprise a varying portion of the ureter from the renal pelvis downward. When the duplication is complete, the ureter from the upper pelvis almost always will enter the bladder more distally (ectopic ureter). In boys, it may open into the trigonum of the bladder, the posterior urethra and, in rare cases, the ejaculatory duct, seminal vesicle or vas deferens. Girls show a corresponding opening into the trigonum, urethra, vulva or the vagina via Gärtner's ducts. It often is not possible to demonstrate radiologically where the ureter has its opening.
Duplication of the ureter often is combined with *ureterocele,* corresponding most often to the ureter from the upper renal pelvis. The ureterocele may be exclusively intravesical or it may extend down through the submucosa into the posterior urethra. It is seen as a filling defect in the contrast material in the lower portion of the bladder. In most cases, the presence of the ureterocele signifies an obstructed flow from the associated renal pelvis, so that hydronephrosis or hydroureter develops. Occasionally, the ureterocele may result in obstruction of the flow through the normal ureter ostium on the same side. Only exceptionally is it possible for a reflux to develop via a ureterocele.

Anomalies of position include inadequate rotation of one or both kidneys, as well as more severe forms, comprising pelvic kidney, usually lying in front of the promontory, and crossed ectopia, where both kidneys lie on the same side of the vertebral column. The ureter from the crossed kidney opens into the bladder normally.

Fetal lobulation has differential diagnostic significance only with respect to pyelonephritic scarring. In lobulation, the indentations are localized between the calices, which are of normal appearance. In pyelonephritis, the indentations are directed toward calices that often are club-shaped.

Monocystic and polycystic degeneration, solitary cysts and *medullary sponge kidney* must all be grouped under congenital anomalies. In particular, the more severe degrees of cystic degeneration, in which one or both kidneys are converted into a conglomerate of large fluid-filled cysts, may occur in infants but are relatively uncommon.

Vascular impressions in the renal pelvis are seen rather frequently. At times, they may result in a hindrance to the flow from the upper calix, almost always on the right side.

Congenital *obstruction of the ureter* generally is localized either to the origin of the ureter from the renal pelvis or to the distal end of the ureter. As a rule, the ureteropelvic stricture causes slowly increasing hydronephrosis. On operation, an aberrant vessel is found at times; in other cases there are only fibrous changes in the wall of the ureter. The ureter usually is not visualized by urography.
Stricture of the lower end of the ureter often is accompanied by a more pronounced hydronephrosis as well as by diffuse dilatation of the ureter down to the bladder wall. The presence of reflux should be excluded before operation.

Retrocaval course of the right ureter may give varying degrees of dilatation down to the level of the 4th or 5th lumbar vertebra, where the ureter is suddenly displaced medially.

BLADDER

Congenital anomalies of the bladder include, among other conditions, complete aplasia and duplication, both very rare anomalies. The latter anomaly may be combined with duplication of the urethra.

Exstrophic bladder is primarily a surgical problem. The radiologic assistance consists of demonstrating possible associated anomalies of the urinary tract.

Other bladder anomalies are complete or partial *persistent urachus*. Diverticula of the bladder may occur at times and should not be mistaken for diverticulous extrusion of the bladder against an inguinal canal.

Neurogenic bladder dysfunction is relatively frequent in children, and may, for example, present as incontinence. It rarely is of traumatic origin, rather a sequel of malformations of the lumbosacral vertebral column (e.g., in myelomeningocele or anorectal atresia), but it often is not possible to demonstrate any cause. In lesions localized above the sacral reflex center of the bladder, a spastic contraction of the external sphincter is obtained, combined with a hypertonic, trabeculated bladder. In addition, diverticula of the bladder often are found. More uncommonly, the cause may be localized to the sacral reflex center or peripheral to it, resulting in a denervated bladder that is flaccid, often large and possibly with overflow incontinence. The diagnosis is made from the radiologic findings on micturition cystourethrography combined with cystometry and urodynamic pressure-flow measurements. Many of these children have vesicoureteric reflux, which probably is secondary to infection of the urinary tract rather than to abnormal intravesical pressure conditions.

URETHRA

Among the urethral anomalies, valves in the *posterior urethra* are the most common cause of infravesical urinary obstruction. Their size and obstructive effect may vary. They cause dilatation of the upper portion of the posterior urethra, in severe cases extending down to the urogenital diaphragm.

Stenosis of the external meatus occurs most frequently among girls. The entire urethra is found dilated, and should not be mistaken for the conical spindle-top urethra, which includes only the upper half of the urethra, and is found as a physiologic condition.

Urethral diverticula occur most frequently in boys and almost always are localized to the anterior urethra. They are best demonstrated during micturition.

Hypospadias and the various *developmental anomalies of the external genitals* in intersexuality are more a clinical than a radiologic problem. Associated malformations are excluded by radiologic investigation, just as the possible presence of a vagina may be demonstrated.

ACQUIRED DISEASES OF THE URINARY TRACT

Urinary tract infection is more frequent in girls than in boys, except under the age of 1 year, where the reverse is the case. The signs — especially in small infants — often are insignificant or uncharacteristic. Clinically, the diagnosis is verified by demonstrating significant leukocyturia or bacteriuria. In the early stages of urinary tract infection there are no or only insignificant radiologic findings in the form of quite slight dilatation of the renal pelvis and/or ureter, unilateral or bilateral. In first-time infections, therefore, radiologic examination of the urinary tract often is omitted. In recurrent infection, urography always should be done, primarily to exclude or demonstrate the presence of congenital or acquired lesions of the urinary tract that might possibly explain the origin of the infection. It should be noted that in many cases it is not possible to demonstrate organic changes. The radiologic examination should also include micturition cystourethrography to reveal any possible reflux of contrast material from bladder to ureter, as well as any possible infravesical obstruction of flow. The presence of *vesicoureteric reflux* is intimately associated with the infection problem. In relatively few cases it is probable that the reflux occurs because of a congenital anomaly of the passage of the ureter through the bladder wall or of the shape of the ureteric orifice, especially in cases in which the reflux is observed in the newborn or in infants. In the majority of cases it is probable that the reflux develops because changes resulting from infection in the wall of the bladder around the intramural course of the ureter hinder normal closing of the distal end of the ureter. Depending on the extent of urinary flow and the degree of distention, the reflux often is classified into different degrees.

If the urinary tract infection becomes chronic, changes gradually appear in one or more calices, which become deformed, and in the renal parenchyma, in the form of *solitary or multiple scars on the surface of the kidney,* together with reduced or interrupted growth of the affected kidney and possibly demonstrable reduction in the renal function. These changes may be due both to the infectious processes in the parenchyma and to a back-pressure effect of the reflux.

In follow-up examinations, an evaluation is made of the growth of the kidney, the thickness of the parenchyma, the development of scars and alterations in the form of the renal pelvis as well as the conditions of flow through the ureter. In all forms of unilateral renal disease and following operative removal of one kidney there usually is an increased growth of the parenchyma in the opposite kidney, a compensatory hypertrophy.

Urinary stones are relatively rare in children, especially in the younger age groups. There is a disposition to the formation of stones in metabolic anomalies, such as renal tubular acidosis, cystinuria and oxalosis, obstructive urinary tract anomalies, pyelonephritis and prolonged immobilization.

Renal tumors are primarily nephroblastoma, *Wilms'* *tumor*, most often diagnosed in the age group 2-3 years but which may also occur in the newborn. They are more uncommon after the age of 5 years. They are bilateral in 5-10 % of the cases. As a result of their size, the shape, size and possibly the position of the kidney are altered, the renal pelvis may be deformed and displaced and the outflow sometimes obstructed. The diagnosis thus may be made by intravenous urography, possibly supplemented by renal arteriography and/or ultrasound scanning.

Renal carcinoma *(hypernephroma)* is extremely rare in childhood.

Renal trauma usually is diagnosed by demonstrating hematuria. In slight trauma, intravenous urography may reveal a normal renal shadow and a contracted renal pelvis, which becomes normal in the course of a few days.

In severe trauma, with rupture of the kidney, it is possible to demonstrate extravasation of contrast material, enlargement of the renal shadow with subcapsular hematoma or a blurred, delimited soft tissue shadow at the site of the kidney, possibly with elimination of the psoas shadow as a result of rupture of the capsule. Renal angiography will provide further information on the localization of the rupture and on the magnitude and extent of vascular lesions. Depending on the nature of the trauma, it may also be necessary to examine whether there is laceration of the ureters, the bladder or possibly the urethra.

Skeleton

Just as is the case in the other organ systems, skeletal changes in children are dominated by a large variety of congenital malformations and developmental disturbances. These may be limited to the skeleton alone, either generalized or involving a single or a few bones, depending on the nature of the malformation, whereas, in other cases, complex malformations are involved, comprising several organ systems. A previous volume, *Roentgen Diagnosis of Bones*, gives a number of examples of these malformations and dysplasias and likewise mentions examples of bone changes of traumatic, endocrine, infectious, hematogenous, vascular, toxic, metabolic and neoplastic nature. At this point, only one further disease complex will be mentioned in which skeletal lesions are included as a significant component – the battered child syndrome.

By the *battered child syndrome* is understood traumatic damage experienced by the child from infuriated and frustrated parents or, in rarer cases, from other persons who are responsible for care of the child. It is seen most often in children under the age of 1 year and rarely after the age of 3-4 years. The syndrome is characterized by the common radiologic demonstration of fractures of several bones at various stages of healing. The lesions that occur most frequently are the result of traction trauma with small avulsions from the metaphyses, epiphysiolyses or dislocations of the limbs. Actual fractures also occur. Fractures of the skull and subdural hematomas are serious and not at all uncommon lesions whereas lesions involving the abdominal organs are relatively uncommon. Also, cutaneous ecchymoses are seen only occasionally. Even in those cases in which the radiologic diagnosis appears obvious, it often is difficult to clarify the causal relationships.

Traction trauma with violent periosteal reaction may also develop during delivery, just as other diseases of the bones, in particular osteomyelitis, may give almost identical radiologic changes.

Radiograms
of Children

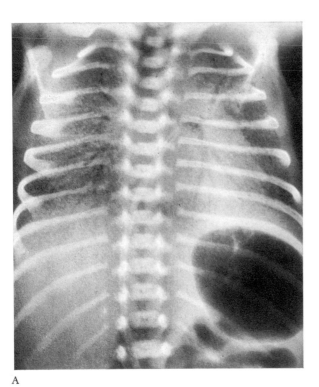

A

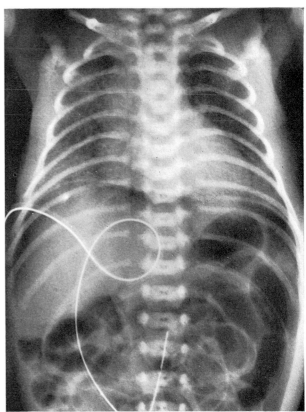

B

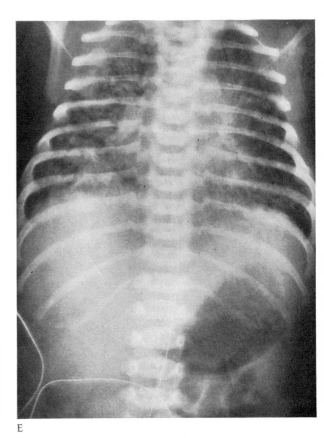

D

E

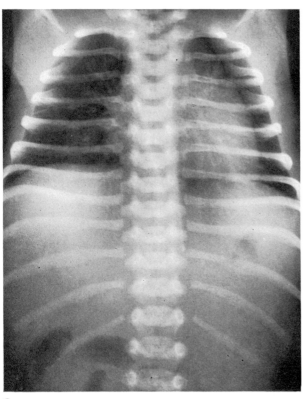

C

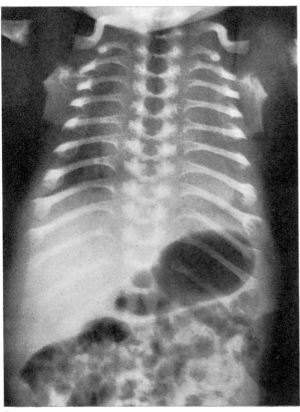

F

Respiratory Distress

The 6 children whose chest radiograms are shown here all have respiratory distress, even though the lung findings appear very different. The reason for the respiratory insufficiency appears, in part, from the films, but the information in the history increases the differential diagnostic possibilities.

A is a boy, born 8 weeks before term (BW 1400 gm, BL 42 cm). Since birth there has been pronounced respiratory distress with cyanosis. The film was taken a few hours after birth.

B also is a newborn boy. He weighed only 1100 gm and immediately from birth had respiratory distress and cyanosis. Heart function was normal. After he was placed in an incubator with 100 % oxygen the cyanosis disappeared, but 24 hours later he was ill with increasing attacks of apnea. Repeated aspiration from the upper respiratory tract brought up bloody mucus. The film was taken within the first 24 hours of life.

C is a newborn girl, delivered at term. Delivery was protracted, made difficult by pronounced edema in the mother, and after birth the infant was pallid, flaccid, with negative reactions, respiratory distress and pronounced cyanosis, alternating with pronounced periods of apnea. There were hypertonic reflexes, tendency to pronation spasm, developing into twitching in the arms and legs, and a few "brain cries."

D is a 3-week-old boy, born 10-12 weeks before term (BW 1240 gm). Immediately after birth he was slightly cyanotic and grunting, so he was placed in an incubator with extra oxygen. During the next 5 weeks attempts were made to stop the extra oxygen, each time resulting in rapid and labored respiration and attacks of cyanosis. Lung stethoscopy was normal and there was no sign of heart disease. Radiograms of the chest during the first 2 weeks after birth showed normal conditions.

E is a girl, born at term. At birth there were excessive amounts of saliva to aspirate and she had generalized cyanosis with rapid respiration and retraction of the thorax along the costal margins. As soon as she was placed in an incubator with 40 % oxygen she quickly developed a good color. After 3 days she was able to do without oxygen and then she did well. The film was taken within the first day of life.

F is a newborn boy with BW 2200 gm, cyanosis and greatly increased rate of respiration. His appearance resembles that of an achondroplastic dwarf with broad skull, bulging forehead, hypertelorism and depressed nasal bridge. The limbs are disproportionately short, with bristling fingers and "trident" hands. Stethoscopy of heart and lungs was normal. He died at 2 days of age.

Make the necessary diagnoses and/or differential diagnoses.

Which of the lung findings are observed particularly in prematures?

Answers

A has *respiratory distress syndrome – hyaline membranes.*

Scattered diffusely in both lung fields may be seen a clearly increased granular (reticulogranular) lung structure, resulting from numerous alveolar atelectases. Against the background of the atelectases, the air-filled bronchial branches may be observed far out into the lung fields – an *air bronchogram*. The patient was placed in an incubator with pure oxygen, followed by intubation and treatment in a respirator, but his condition worsened and he died at 1 day of age.

B has *pneumoperitoneum*. The film was made with the patient supine in an incubator, and the air therefore is distributed over the whole abdomen, most clearly along the diaphragm and the left flank. A supplementary exposure in the supine position with the x-ray beam horizontal (B-2) shows a large amount of free air under the anterior part of the diaphragm.

Just as with large space-occupying lesions in the abdomen, the presence of a large amount of free air in the peritoneum may inhibit the movement of the diaphragm and cause respiratory distress. In this patient, however, small patchy consolidations are seen basally in both lungs, most clearly on the right side; otherwise, the lungs appear to be normally expanded. The changes are nonspecific, but repeated aspirations of bloody secretion could suggest *pulmonary hemorrhage,* a not uncommon finding in premature infants, especially in hypoxia and often together with hyaline membranes. This infant died at 2 days of age. Autopsy showed violent hemorrhages in the lung alveoli but no hyaline membranes. There also was bleeding into the ventricles of the brain. The cause of the pneumoperitoneum was not demonstrated. The most important causes of this in the newborn are perforation of the stomach or perforation of the intestine in necrotizing enterocolitis or intestinal obstruction, such as atresia, meconium ileus or volvulus. Finally, in pneumomediastinum, it should be possible for the air to spread to both the retroperitoneal connective tissue and the peritoneal cavity.

C shows well expanded lungs. The combination of *normal lungs* and respiratory distress, especially when the latter is characterized by periods of apnea, suggests brain damage. *Intracranial hemorrhage* – subdural, subarachnoidal, intraventricular or in the brain parenchyma – represents one of the most serious birth complications, with high primary lethality and risk of permanent brain damage. In *C*, a radiogram of the skull showed a fracture through the occipital bone *(C-2)*. However, it is exceptional that radiologic examination of the skull will reveal abnormal conditions. After treatment with oxygen for 2 days the patient improved. Two months later, the EEG was normal, and a follow-up examination at the age of 1½ years likewise showed completely normal conditions.

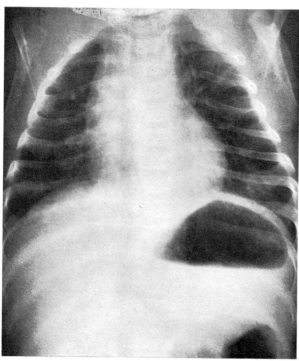

B-2

D-2

D had very modest signs at birth, but frank cyanosis and respiratory distress developed in the course of the first week. Even during the period following withdrawal of oxygen treatment, when he was more than 5 weeks of age, he had several attacks of slight cyanosis and rapid respiration. The radiogram shows a coarse reticular network of streaky densities in both lungs. The clinical picture and the pulmonary findings are characteristic of the *Wilson-Mikity syndrome,* and the coarse reticular structure is due to an increased amount of interstitial tissue. The Wilson-Mikity syndrome is found almost exclusively in premature infants who usually are born at least 4 weeks before term. The clinical signs disappear earliest after 4-6 weeks but often persist for many months. The radiologic findings persist longer than the clinical ones, with decreasing intensity. D-2 is a control film taken when the patient was more than 3 months of age. Remains of the increased interstitial markings still can just be seen.

E has patchy consolidations scattered throughout both lungs, suggesting small atelectases, as is seen to varying degrees in *the aspiration syndrome* and *wet lungs.* The appearance of the consolidations may vary from modest linear atelectases fanning out from both hili to massive segmental atelectases. Both the pulmonary findings and the respiratory distress disappear most often in the course of hours to a few days. However, the changes are nonspecific, and as it was not possible to exclude bronchopneumonia following aspiration of infected amniotic fluid, the patient was treated with antibiotics. The patient was free from symptoms in the course of 3 days, and the pulmonary picture normalized *(E-2).*

F has a very narrow thorax with short, coarse ribs and very flat, incompletely ossified vertebral bodies. Together with the physical findings, these changes correspond to the picture of *thanatophoric dwarfism.* Radiologically, this is characterized by changes qualitatively identical but quantitatively more severe than those in achondroplasia. The narrow thorax is characteristic of thanatophoric dwarfism. In contrast to achondroplastic dwarfs, thanatophoric dwarfs almost all die in the course of a few hours to days after birth.

Deformity of the thoracic cavity, also seen, for example, in Jeune's asphyxiating thoracic dysplasia and in osteogenesis imperfecta with multiple costal fractures, often results in respiratory distress.

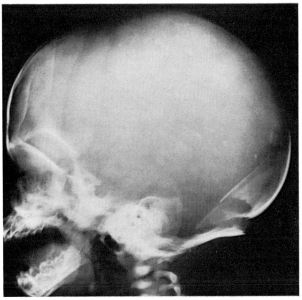

C-2

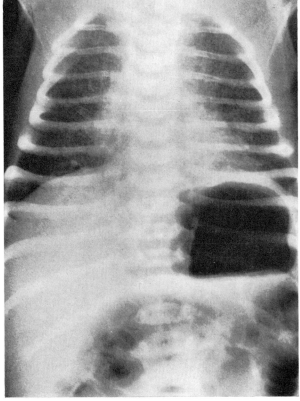

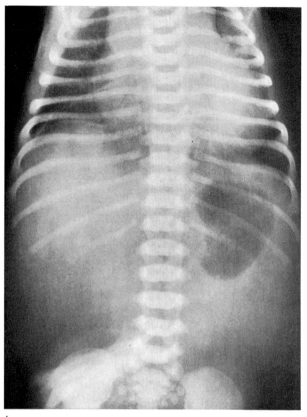

A

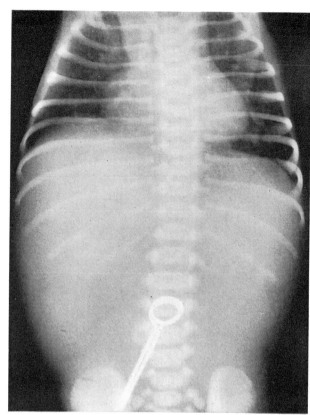

B

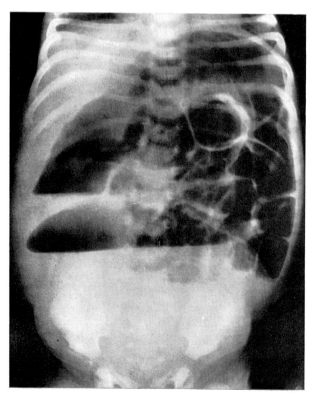

D

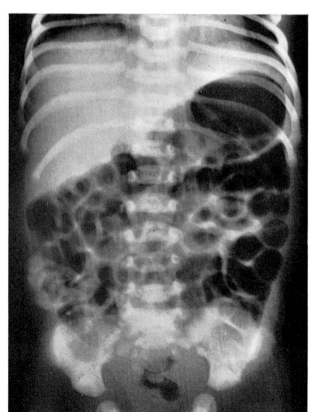

E

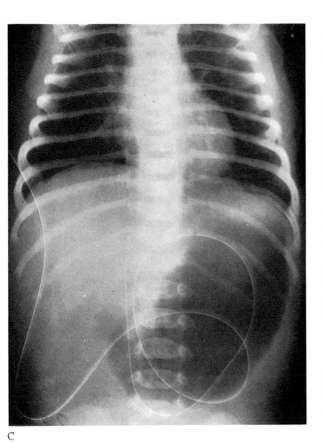

C

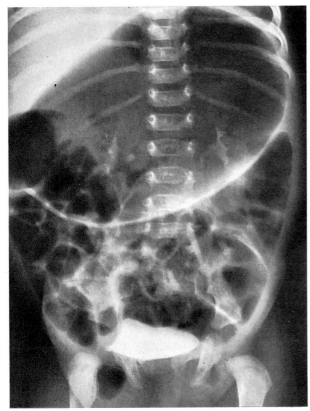

F

Abnormal Distribution of Air in the Digestive Tract

In the healthy newborn infant there almost always is air in the gastrointestinal tract immediately after birth. Most of this air reaches the stomach by swallowing movements whereas respiratory movements are thought to play only a subordinate role. After 1/2-1 hour, the air usually has reached the small intestine, after 3-4 hours it has reached the cecum and after 9-10 hours the sigmoid colon.

The physiologic variations in the distribution to the various intestinal segments are rather great, and even a delay of several hours is not necessarily an expression of intestinal obstruction.

In the affected newborn, for example with the respiratory distress syndrome, air already present in the intestinal tract may disappear completely or partly, probably because of a displacement in the child's fluid balance in the direction of dehydration.

A, B and C have no or strikingly little air in the intestinal tract whereas D, E and F have plentiful intestinal air. F's film is from intravenous urography. The pronounced distention of the stomach results from the patient having been given a bottle with his usual mixture after the contrast material was injected. This results in distention of the stomach, pushing the intestinal air distally, so that the kidney shadows and the renal pelves are projected completely or partly within the stomach air.

Compare the films with the following clinical information.

A is a boy, born 2-3 weeks before expected time. Labor was prolonged but otherwise uncomplicated. He cried healthily at birth. Respiration is normal and color is good. The film was taken 1 1/2 hours after birth.

B is a newborn boy, 2 weeks premature and admitted for anal atresia. Excessive frothy secretion is seen around the mouth and plentiful amounts are aspirated from nose and throat.

C is a 3-day-old girl, born at term. Since birth she has regurgitated, had bile-stained vomitus and is admitted in a dehydrated condition. She has passed meconium and the abdomen is soft, not distended.

D is a 6-week-old boy, born 2 weeks after term. When he was 2 weeks of age, the abdomen was found distended, but after laxatives plentiful amounts of meconium were passed. He now is admitted because of an increasing distention of the abdomen and signs of ileus (see radiogram). A rectal tube was introduced and large amounts of fetid, thin feces and air were removed.

E is a 1-day-old boy, born 2 weeks before term and admitted for prematurity. Respiration is normal and there are no signs from the gastrointestinal tract.

F is a 16-day-old boy, admitted with pyuria. Intravenous urography was performed. He has never had any symptoms from the intestinal tract and the abdomen is soft.

Compare the radiologic findings with the clinical information. What is the matter with these children?

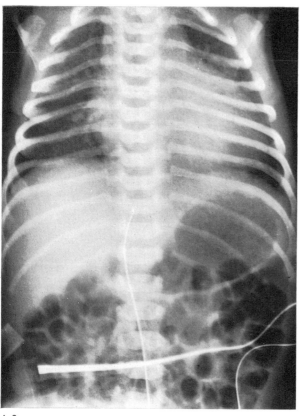

A-2

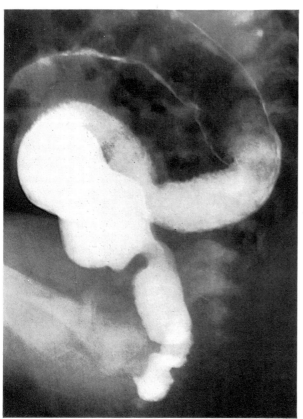

D-2

Answers

A's film actually is a chest radiogram, but first let us look at the abdomen. This is striking because there is only a little bubble of air in the stomach, and no intestinal air. Without knowing the patient's age, we are inclined to suspect duodenal obstruction. However, only 90 minutes have elapsed from birth to examination, and the findings thus are not *definitely pathologic*. Clinically, the infant was unaffected and a follow-up exposure on the next day *(A-2)* showed completely normal distribution of air throughout the entire intestinal tract.

The heart seems slighty enlarged, as often is the case on films taken immediately after birth. Also, this was normalized in the course of 24 hours.

B has an accumulation of malformations. The film, which was taken several hours after birth, shows no air in the gastrointestinal tract. The frothy secretion around the mouth makes us suspect *esophageal atresia*, which was confirmed. *B* has esophageal atresia *without a fistula from trachea to lower esophageal segment*, and the air therefore has no possibility of passing down into the stomach. Such atresia constitutes less than 10 % of all esophageal atresia. Gastrostomy and colostomy were performed. At operation there were also demonstrated atresia of the duodenum at the level of the papilla of Vater, atresia of the gallbladder and of the cystic duct, as well as malrotation of the intestine. The patient died at 11 days of age.

C has only air in the stomach and duodenal bulb. She is 3 days of age, and the radiogram and clinical examination together indicate the presence of *duodenal obstruction*. A small amount of barium was injected into the stomach through a tube. *C-2* shows how most of the contrast material still is retained in the stomach and duodenum, but small patches of barium are seen immediately below the duodenal obstruction and in the vicinity of the left iliac bone. *C* has a thin *septum in the pars descendens duodeni* with a quite small opening. At operation, malrotation of the intestinal tract was also demonstrated.

D's history, compared with the abdominal scout film showing severe gaseous distention, mainly of the colon, with a broad air-fluid level, arouses suspicion of *congenital megacolon (Hirschsprung's disease)*. Two days later a barium enema was done. *D-2* shows a lateral film, taken 24 hours after the enema. There is a moderate disproportion between the caliber of the upper and lower part of the rectum, suggesting *distal aganglionosis*. Anal tonometry showed pressure conditions typical of Hirschsprung's disease. As the boy was emaciated and in poor condition, sigmoidostomy was performed. One year later, the rectum and sigmoid were removed up to the sigmoidostomy. The remaining part of the sigmoid was anastomosed to the posterior wall of the rectum, just above the anus (Duhamel operation). Microscopy of the stenosed part of the rectum showed absence of ganglion cells.

The patient has been followed up for a further 17 months. Apart from an isolated ileus-like attack, he has managed well.

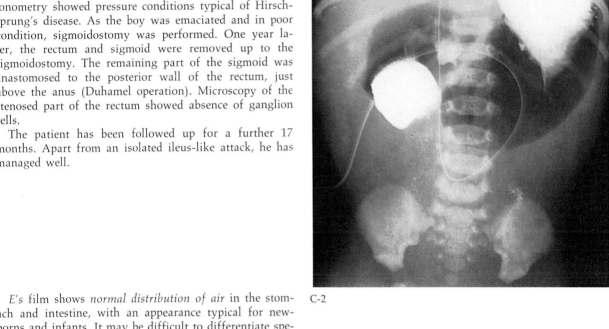

C-2

E's film shows *normal distribution of air* in the stomach and intestine, with an appearance typical for newborns and infants. It may be difficult to differentiate specifically colonic air except for its presence in the rectum. Likewise, there are no cardiopulmonary changes.

Intestinal air usually is observed as small, closely lying bubbles, but varying degrees of distention of minor segments of the intestine often are seen. Scattered, small air-fluid levels may also be physiologic in this age group. A demonstration of a pathologic state therefore must be based on a combined clinical-radiologic evaluation.

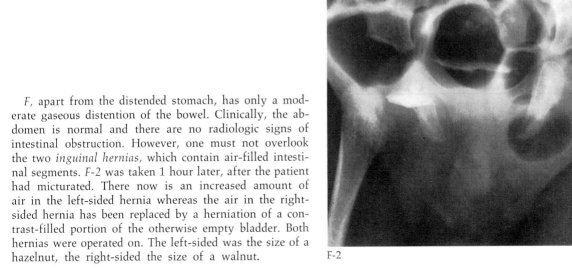

F, apart from the distended stomach, has only a moderate gaseous distention of the bowel. Clinically, the abdomen is normal and there are no radiologic signs of intestinal obstruction. However, one must not overlook the two *inguinal hernias*, which contain air-filled intestinal segments. *F-2* was taken 1 hour later, after the patient had micturated. There now is an increased amount of air in the left-sided hernia whereas the air in the right-sided hernia has been replaced by a herniation of a contrast-filled portion of the otherwise empty bladder. Both hernias were operated on. The left-sided was the size of a hazelnut, the right-sided the size of a walnut.

F-2

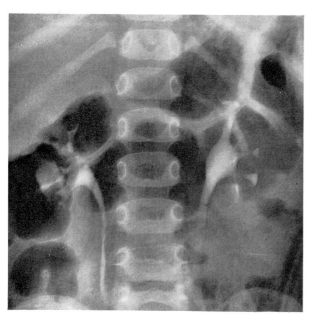

A

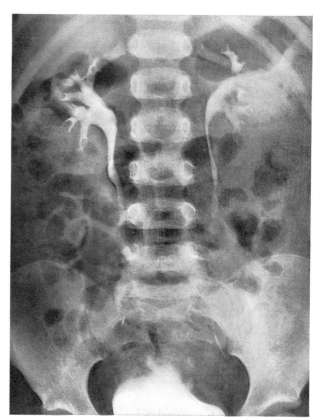

B

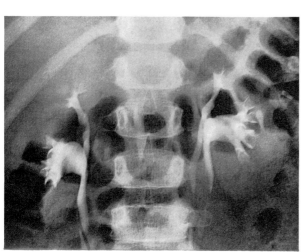

D

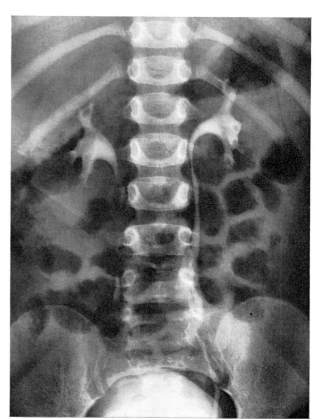

E

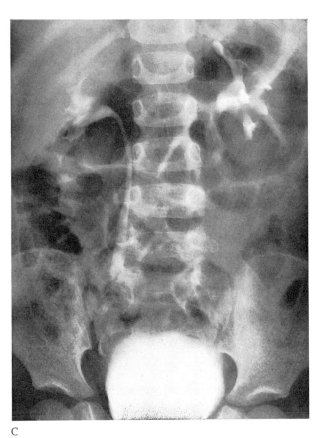

C

F

Urinary Tract Infection

Urinary tract infection in children has long been in the limelight, but despite both improved and completely new methods of investigation, improved techniques of operation and comprehensive follow-up of the patient, numerous problems still remain unclarified.

All of these patients have been hospitalized for recurrent urinary tract infection, but that is also the only thing they have in common. Here are their histories:

A is a 21-month-old girl. In the course of the past 12 months she has been hospitalized 4 times for urinary tract infection and has been under almost constant chemotherapy.

B is a 3½-year-old girl. In the course of the past 6 months she has had 2 attacks of urinary tract infection and she now is hospitalized subfebrile with a renewed infection.

C is a 3-year-old girl who was hospitalized 6 months ago for recurrent urinary infections. An intravenous urogram showed only that the right renal pelvis was slightly larger than the left, but micturition cystourethrography (MCU) demonstrated severe reflux to both ureters and renal pelves, for which reason a bilateral antireflux operation was performed using the method of Leadbetter-Politano. There has been no urinary tract infection since, and she now is hospitalized for a follow-up examination.

D is a 5-year-old girl admitted for "recurrent pyuria". During the past 2 years on several occasions she has had irritation and reddening of the skin of the labia, and occasionally there has been nocturnal enuresis. It is stated that pyuria was demonstrated before hospitalization, but, during this, all urinary tract examinations, just like other laboratory investigations, have in fact been normal.

E is a 3-year-old girl, hospitalized with fever of 1 week's duration and pain in the right side of the abdomen. During the past 24 hours she has had a single attack of vomiting and diarrhea. There has been no urinary tract infection previously, but pyuria and significant bacteriuria were demonstrated on hospitalization.

F is an 8-month-old boy, hospitalized with massive pyuria. At the age of 1 month he was hospitalized with acute gastroenteritis, with vomiting and thin bowel movements. On that occasion, albuminuria and pyuria were demonstrated, which disappeared on chemotherapy but recurred on terminating the therapy.

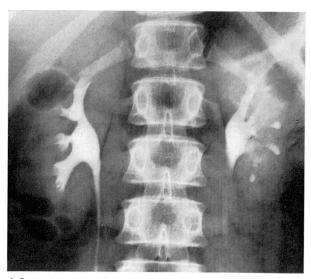

A-2

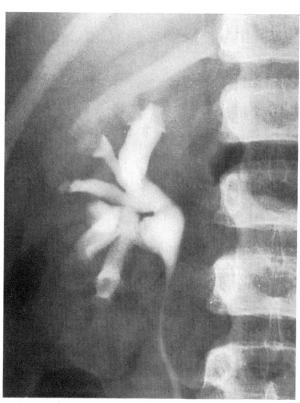

B-2

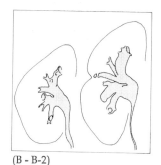

(B - B-2)

Tracing of the right Kidney with a 3-year interval.

Answers

A has a *very slight dilatation of the left renal pelvis and ureter,* but micturition cystourethrography 10 months ago showed moderate reflux to both ureters, most pronounced on the left side. A renewed MCU shows that the reflux on the right side now has disappeared, but there is increased reflux on the left side. The long-term prophylactic chemotherapy that had been started therefore was continued for a further 6 months, without the patient having a further relapse.

We now will jump 11 years ahead in time. She has not had urinary tract infection since those attacks already referred to, but she still has moderate reflux on the left side. Intravenous urography *(A-2)* reveals that the left kidney is 2 cm shorter than the right (11 cm and 13 cm, respectively) and that the calices are club-shaped. The contour of the kidney is irregular, and the lateral parenchyma measures a maximum of 10 mm compared to 25 mm on the right side. The findings are supposed to be the result of *pyelonephritic scarring, back-pressure atrophy resulting from the reflux* and a transitory restrained growth. On isotope scintigraphy, the function of the left kidney was found to be only one-third that of the total renal function. With the aim of stopping further atrophy, a left-sided antireflux operation was performed using the method of Leadbetter-Politano.

B's urography shows that the *right renal pelvis is slightly better filled than the left.* Both renal shadows are normal. However, MCU showed bilateral reflux with filling of the renal pelves but only very slight dilatation. Despite prolonged chemotherapeutic treatment, she had recurrent urinary tract infections during the subsequent 3 years, and radiologic investigations revealed the development of a *solitary scar* at the middle of the lateral contour of the right kidney, stretching of the underlying calix and a small patch of contrast material lateral to this, probably in a former abscess cavity of the papillary apex *(B-2)*. Isotope scintigraphy was normal, but as reflux still was present, bilateral neoureterocystostomy was performed. During the course of a further 7 years' observation, she has had no urinary tract infection, and the radiologic picture has remained unchanged.

C's urography shows a left kidney measuring 7 cm compared to 8 cm for the right kidney. *The parenchyma of the left kidney measures only 8-9 mm compared to 20 mm on the right side.* Most of the left-sided calices are deformed, with flattening of the papillae. C-2 shows the degree of the reflux before the operation 6 months previously. Atrophy, scarring and restrained growth is the supposed cause as in case *A.*

This girl is now almost 7 years old. In the past 3 years there have been no urinary infections. The radiologic changes in the left kidney have been stationary. However, there is no reflux and the left kidney is growing, so that its long axis now is 7,6 cm compared to 8,8 cm for the right kidney. Isotope scintigraphy shows that the left kidney function amounts to only slightly more than one-third of the total renal function.

D has a *developmental anomaly* of the urinary tract, both renal pelves and ureters being duplicated. Such duplications may occur unilaterally or bilaterally. The two ureters on each side may coalesce at any point in their passage to the bladder or may remain separate. In *D*, the ureters opened into the bladder in close pairs.

In renal duplications, the upper pelvis almost always will be the smaller. The condition does not represent a doubling of the kidney but a splitting up of the renal pelvis.

When the ureters in doubling have a separate course, when they open independently into the bladder and when there is no obstruction of the urinary flow, the abnormality has hardly any significance for the development of a possible urinary tract infection.

If the two ureters of a double structure coalesce before reaching the bladder, peristalsis in the two ureters will not be synchronous, so that urine from the one ureter may flow up into the other ureter. This is a kind of reflux, and like all other forms of compromised urinary flow, it may have etiologic significance for the development of an infection. It is doubtful whether *D* has ever had a urinary tract infection, as this has never been verified. It is more likely that the irritation of the genitals has been due to her enuresis.

E's intravenous urography is completely *normal*. She received chemotherapy for 4 weeks. One year later, urinary tract infection was demonstrated again, and urography now showed pronounced deformation of the right-sided calices and slight dilatation of the right renal pelvis. MCU showed insignificant reflux to the lower ends of both ureters, not definitely pathologic. She since has been under almost continuous treatment with antibiotics and chemotherapy. The urine has been sterile during the treatment, but each time the treatment was withdrawn she had a relapse. The changes in the right kidney have gradually increased, the kidney has *stopped growing* and *severe reduction of the parenchyma* has developed, but the reflux has disappeared. *E-2* shows the conditions on intravenous urography 7 years after the first hospitalization. The right kidney now measures only 7,5 cm compared to 12 cm for the left kidney. Reduction of the parenchyma is most pronounced in the region of the upper right pole, where the parenchyma measures 5 mm, in contrast to 33 mm at the left upper pole. There is compensatory hypertrophy of the left kidney, which otherwise is normal.

F has severe *deformation of both renal pelves*, with irregular deformity of all calices. The ureters are dilated and the middle third of the right ureter shows pronounced *longitudinal linear striation*, presumably due to intermittent distension, and mainly seen in patients with reflux. MCU showed heavy reflux to both ureters and pelves. The urine could be kept sterile during prolonged prophylactic chemotherapy, but the pyuria recurred as soon as the treatment was withdrawn. Seven months later, an antireflux operation was performed by the method of Leadbetter-Politano, with a good result. The patient was followed up for a further 7 years. The infections have disappeared and both kidneys show satisfactory growth.

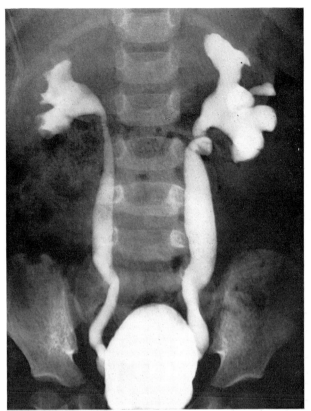

C-2

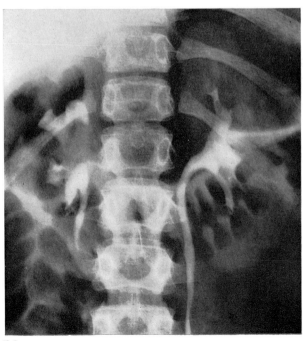

E-2

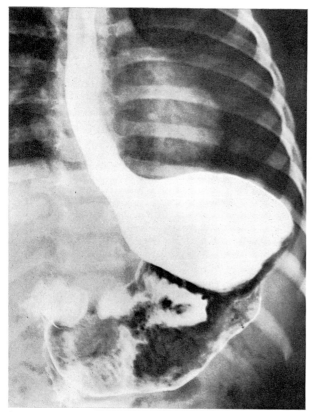

A

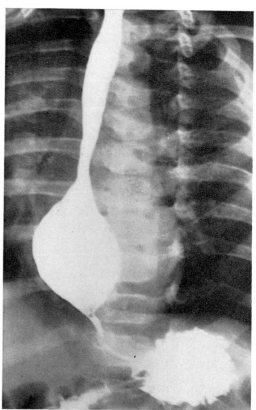

B

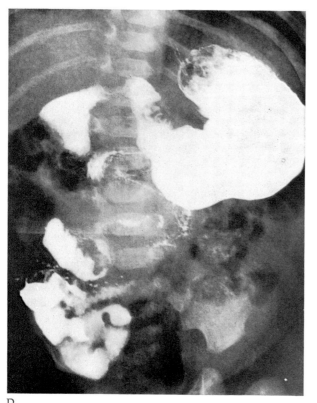

D

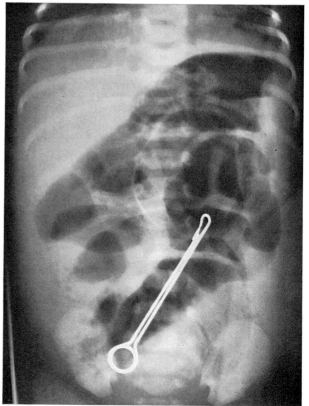

E

Vomiting of Gastrointestinal Origin

Vomiting is a very common sign in children, not only in diseases of the gastrointestinal tract but also as a more or less nonspecific reaction to many other diseases.

The patients on these pages all have attacks of vomiting due to changes in the gastrointestinal tract.

A is an 11-month-old girl, admitted for failure to thrive and for numerous minor attacks of vomiting since birth. Radiologic examination of the esophagus and stomach shows frequent and plentiful gastroesophageal reflux.

B is a 5-month-old boy. From the age of 2 months he has had several minor attacks of vomiting during almost all meals.

C is a 6-week-old boy who has had 1-2 small attacks of vomiting daily since birth. During the past week, the episodes have become more frequent, appearing after every meal, and often are projectile. Gastric peristalsis is observed. The film of the stomach was taken with the patient lying on his right side.

D is a 6-week-old boy who has been vomiting after nearly every meal for 1 week. The vomiting often is projectile and gastric peristalsis is clearly seen. The film was taken with the patient lying supine, 30 minutes after a barium meal.

E is a newborn girl. Three of 4 siblings have died from cystic fibrosis of the pancreas. When 14 hours old, her abdomen started to be tense, distended and with increasing venous markings. There was no passage of meconium.

F is a 4-year-old girl, admitted with acute abdomen. For a week she has been complaining of attacks of colicky pain in the vicinity of the umbilicus, accompanied at times by non-bile-stained vomiting. On hospitalization, a 6 x 4 cm, well-delimited, movable mass can be felt below the left costal margin. She is seen to pull hair out and eat it, and it is stated that for some time she has been swallowing indigestible objects such as bits of paper, knitting wool and foam rubber. Laboratory investigations are normal. Radiologic examination of the stomach is performed.

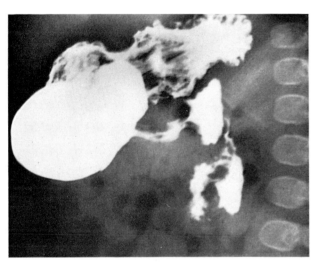

C

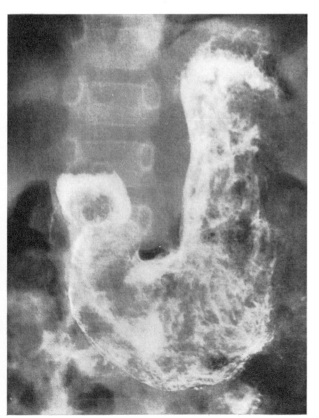

F

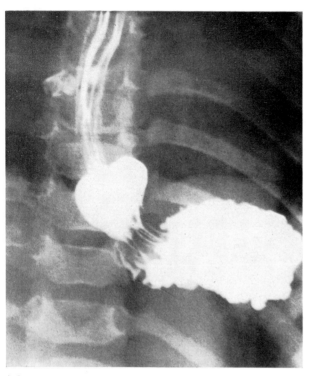

A-2

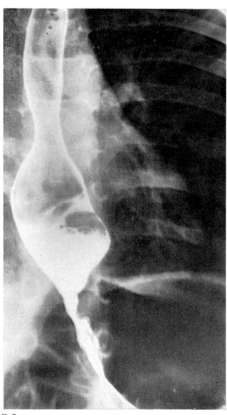

B-2

Answers

A apparently is suffering from a simple cardioesophageal incompetence – *chalasia* – since the barium runs back from the stomach to the esophagus almost unhindered when the patient is placed in the supine position. However, further investigation shows that a *small sliding hiatal hernia* is present. Such hernias are the most frequent cause of passive reflux of the gastric contents to the esophagus.

B has cardiospasm – *achalasia of the esophagus*. The greatly obstructed passage through the cardia has caused a dilatation of the lower portion of the esophagus. The patient underwent treatment by myotomy using Heller's method, in which the musculature in the distal part of the esophagus and corresponding to the cardia is split longitudinally, by analogy with pylorotomy in pyloric stenosis. However, the results of the operation are not always equally good in the long term. In some of the patients, the cardiospasm is reconstituted after a transient improvement and the esophageal dilatation continues, sometimes with surprisingly few subjective symptoms. These patients, therefore, always should be followed up with radiologic control for several years after the operation. *B-2* shows the esophagus in this patient 3 years after the operation.

C's problem is localized to the pyloric canal. He has *infantile (congenital) hypertrophic pyloric stenosis*. The circular pylorus muscle is greatly enlarged and obstructs the lumen. This state is the most frequent cause of surgical intervention on the digestive tract in infants. The radiologic examination is best carried out with the infant placed on his right side, possibly turned a little to the prone position. In this way, optimal filling of the distal portion of the stomach is achieved, and by small rotations of the patient the elongated and stenosed pyloric canal can be projected free of the stomach. The patient was treated by pylorotomy *(Ramstedt's operation)*, in which the hypertrophic musculature is split longitudinally right down to the mucosa. During the first 8-10 days after the operation, a number of the children continue to have attacks of vomiting. On radiologic examination during this period it is not unusual for the pyloric canal to still appear abnormally long and narrow, but if the stomach emptying is observed for 15-30 minutes, it takes place at more or less the normal rate. The postoperative findings probably are due to edema and slight hemorrhage in the region.

D's film is more difficult to interpret and provides only incomplete diagnostic information. There is a pronounced obstruction of the third portion of the duodenum and the descending portion is displaced to the right. Both findings could be due to a space-occupying lesion in or near the head of the pancreas; for example, a pancreatic or mesenteric cyst. Gastrointestinal duplication is also a diagnostic possibility. By following the passage of the barium meal through the small intestine, the loops of jejunum were found in the right side of the abdomen, indicating the presence of an *anomaly of rotation*. A barium enema *(D-2)* shows the cecum lying in front of the third portion of the duodenum. *D* has *nonrotation of the intestine with secondary rotation of the ileocecal region* to the right.

Anomalies of rotation appear in fetal life when the midgut – comprising the last portion of the duodenum, the jejunum and ileum and the proximal half of the colon – returns from the umbilical sac into the abdomen. This return normally is associated with a rotational movement of the midgut around the axis of the superior mesenteric artery, but in some cases the movement fails to appear, is incomplete or is reversed.

In addition to the typical nonrotation (mesenterium ileocolicum commune), numerous variations are observed on radiologic examination. Some of these occur as a result of the midgut performing secondary rotations clockwise or counterclockwise (malrotation), so that the cecum may be seen in various positions.

Peritoneal bands often stretch from the abnormally placed cecum and ascending colon up toward the right side of the abdomen, overlying and eventually compressing the duodenum. This may cause varying degrees of obstruction, which, in severe cases, cannot be distinguished from duodenal atresia. Volvulus of the small intestine is not uncommon as the midgut rotates around the narrow mesenteric stalk. In *Ladd's operation*, the peritoneal band is transected, a possible volvulus is untwisted and the cecum and ascending colon immobilized in the left side of the abdomen, as in nonrotation.

D had no volvulus, but otherwise the operation was carried out as described, and he has not had symptoms in the subsequent 5-year period of observation.

E has ileus with strongly dilated, air-filled loops of small intestine. This could be caused by an atresia, but in view of the history, *meconium ileus* is probable. Operation 24 hours after birth showed large masses of meconium accumulated in the lowest part of the ileum. A 25-cm-long portion of grossly dilated ileum was removed and the remainder of the intestine was flushed with saline. Postoperatively, the patient received treatment with antibiotics and pancreatic enzymes, but gradually deteriorated and died 6 days later. At autopsy, the intestinal tract was found to be patent. Bronchopneumonia, a patent foramen ovale and ductus Botalli were demonstrated. Histologic examination of the pancreatic tissue showed cystic fibrosis.

Resection of the intestine rarely is necessary. Through a double-barreled ileostomy, the small bowel may be flushed intermittently or continuously over a period of 1-2 weeks, until the passage is free, after which the ileostomy is closed.

F has a large filling defect in the barium shadow of the stomach. From the history, it is exceedingly probable that it is produced by a bezoar, a mass constituted of hair or other indigestible components that the patient has swallowed. The patient was operated on and a *trichobezoar* the size of a hen's egg removed.

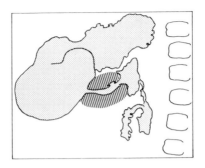

(C)
The hypertrophic pylorus muscle is seen hatched.

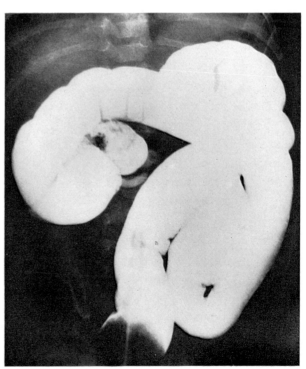

D-2

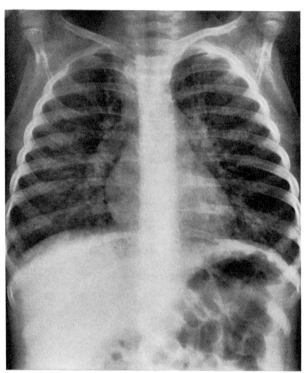

A

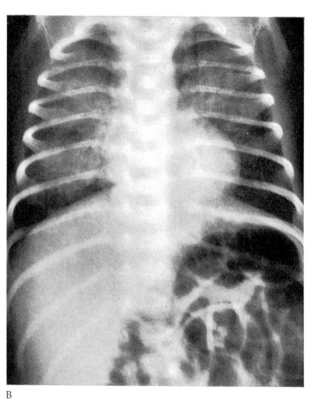

B

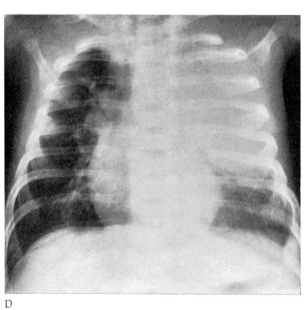

D

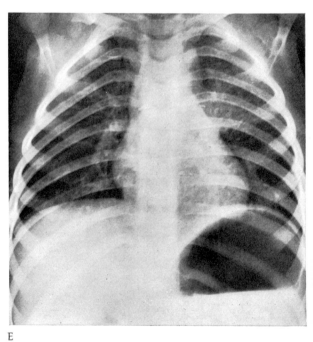

E

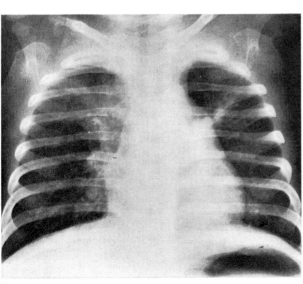

C

F

Pulmonary/Pleural Infections

Inflammatory processes in the lungs may result in very varied radiologic findings. These findings may be due either to the occurrence of inflammatory alveolar exudate, which dominates in alveolar pneumonia (bronchopneumonia, segmental and lobar pneumonia), or to inflammatory changes in the walls of the bronchi and in the peribronchial and interalveolar interstitial tissue in interstitial pneumonia. Secondary changes may develop in the form of atelectasis-like consolidation or emphysema of greater or smaller pulmonary segments, abscess formation, bronchiectasis or pneumatocele. If the infection spreads to the pleura, pleural thickening or massive pleural effusions are seen.

Three of the patients on these pages have acute respiratory tract infection, a further 2 have chronic infection whereas 1 has pulmonary changes of a noninfectious nature. Try to classify the patients into these 3 groups before you read the following information.

A is a 2-year-old girl, previously well. During the week prior to admission she has had a cold, with poor appetite, cough and fluctuating temperature. Penicillin treatment was started 5 days before hospitalization, but without effect. On hospitalization, she had a high fever (39.8° C) with slight cyanosis, so she was placed in an oxygen tent.

B is a 21-day-old girl, born 4-5 weeks before term, somewhat premature but otherwise completely satisfactory. Immediately before the radiologic examination, during a meal, she developed a poor color, respiration became squeaky and there was a reduction in breath sounds on both sides.

C is an 8-month-old girl. During the first few weeks after birth she had repeated foul-smelling, fatty stools, often 3-4 times daily. She improved on a fat-free diet, but at the age of 6 months she started to have recurrent asthma-like symptoms with cough and fever. Radiologic examination of the lungs showed bilateral bronchopneumonic consolidations.

D is an 8-week-old girl, admitted for pneumonia. Four days previously she suddenly developed a high fever, sore throat, cough and tachypnea. Penicillin treatment was without effect. On hospitalization, her cerebral function appeared affected, but examination of the cerebrospinal fluid showed nothing abnormal.

E is a 2½-year-old girl, living under poor social conditions. She has been tired for a week, irritable and with a slight cough. On hospitalization, she has a high fever but is otherwise without symptoms.

F is a 7-month-old boy who has had a severe cold for 2 days, with high fever, wheezy respiration, cough and tachypnea.

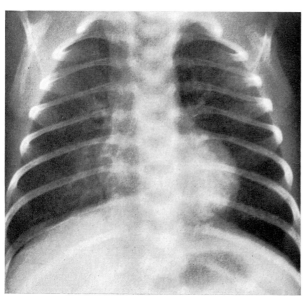

A-2

Answers

A has *interstitial pneumonia,* probably *viral.* The hilar shadows are enlarged, the bronchovascular markings are increased and here and there scattered small consolidations are seen. Without knowing the history, the picture could arouse suspicion of cystic fibrosis of the pancreas (cf. *C-2),* but with continued antibiotic therapy the pneumonic changes disappeared and a follow-up examination 1 week later *(A-2)* showed the lungs to be normal.

B has widely scattered, partially confluent consolidations in both lungs. It may be bronchopneumonia, but the history suggests *aspiration of food.*

Aspiration of food occurs particularly during the first months of life, mainly in prematures with immature swallowing reflexes. The condition often develops into bronchopneumonia. The patient was treated by suctioning, physiotherapy, oxygen and penicillin, and 1 week later the clinical and radiologic findings had disappeared *(B-2).*

C has *cystic fibrosis of the pancreas.* Failure to put on weight and the abnormal stools aroused a suspicion of this condition, but it was only when she was 7 months of age that the diagnosis was confirmed by the demonstration of an abnormal sweat test.

Cystic fibrosis of the pancreas is a recessive hereditary disease, localized to the eccrine sweat glands and the mucous glands. The disease presents in about 10 % of patients with meconium ileus, but it did not appear in the present patient. As to the sweat glands, it appears as elevated concentration of sodium and chloride in the sweat. Blockage of the glandular ducts from the pancreas produces degenerative changes with reduction in the external pancreatic function, resulting in steatorr-

B-2

C-2

hea, loss of weight and inhibition of growth. The liver function may also be affected. In the lungs, the bronchioles are plugged by viscid mucus, resulting in recurrent infections and bronchiectases. The radiologic findings in the early stages are those of an acute pneumonia, but in the course of the disease, chronic changes develop, with progressive peribronchial infiltration and interstitial fibrosis, combined with varying degrees of localized obstructive emphysema and segmental atelectasis. C-2 shows the lungs when the patient was 8 years of age. The film was taken when the disease was in a quiet phase. There are widespread chronic interstitial changes, combined with stagnating bronchial secretion and strongly accentuated hilar shadows. Despite antibiotic treatment of the pulmonary infections, supported by pulmonary physiotherapy and dietetic and medicamental treatment of the pancreatic insufficiency, the patient died at the age of 14 years with signs of increasing pulmonary insufficiency and terminal development of cor pulmonale.

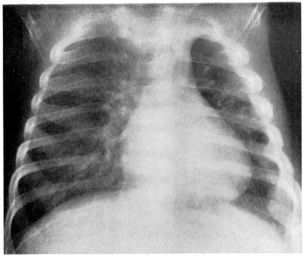

D-2

D has diffuse shadowing of the entire left lung, most intense above. A pronounced pleural exudate is seen along the lateral thoracic wall. The exudate extends over the anterior and posterior surface of the lung, thus causing the diffuse blurring. Culture from tracheal secretion showed staphylococci. Despite continued antibiotic therapy, the shadow increased in density during the next 4 days, so a rubber tube was placed in the pleura. Pus was removed, and this also showed growth of staphylococci. *Empyema pleurae* develops in most such cases as a complication of acute or chronic pneumonia in infants and young children, most often caused by staphylococci, streptococci and *Hemophilus influenzae*, often early in the course of the disease. The inflammatory changes in the lungs usually are not visible until after drainage of the pleural exudate.

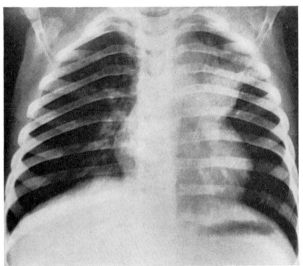

E-2

E has *primary pulmonary tuberculosis*. A few days previously, a younger brother was hospitalized with the same disease. Because of perifocal alveolar exudate, it is not possible to see a typical primary complex, but an ill-defined shadow is seen extending from the upper portion of the left hilum. However, there was a strong positive tuberculin reaction, and culture of gastric washings evidenced growth of numerous colonies of human tubercle bacillus. By the time the culture results were available, antituberculous treatment had already been started. Despite the treatment, the radiologic changes progressed during the first month, as the perifocal shadow increased in density and obscured a larger portion of the lung *(E-2)*. The shadow then cleared slowly and after 5 months' treatment the patient could be regarded as noninfectious.

F has *bronchopneumonia* with mainly perihilar spread in the right lung. In the apex of the left lung a linear density is seen, probably a small atelectasis due to excessive bronchial secretion. After penicillin treatment, the temperature was rapidly reduced and the patient could be discharged 5 days later. On the lateral view, the consolidation is seen extending upward and backward, but, in addition, there are small patches of densities in the region of the lower lobe.

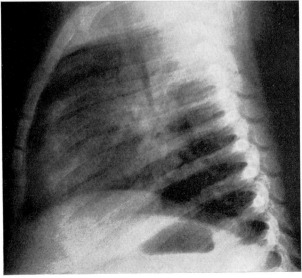

F-2

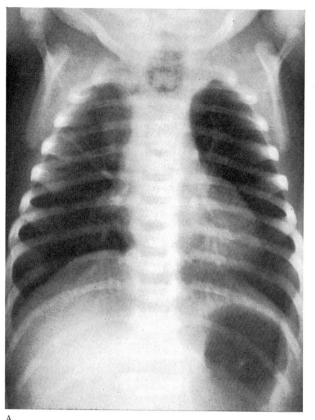

A

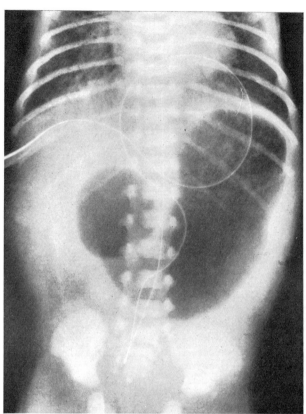

B

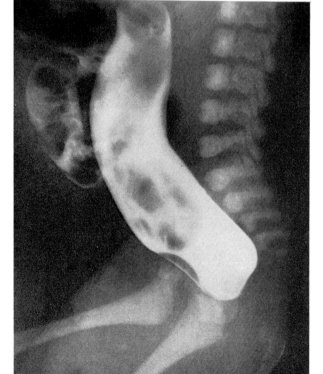

D

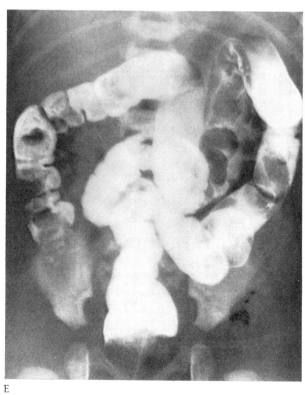

E

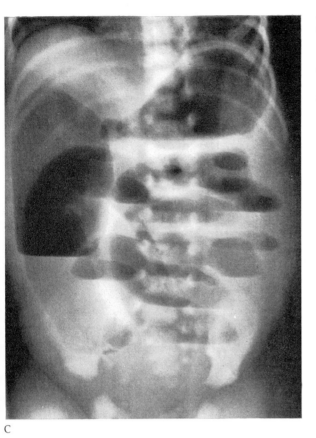

C

F

Intestinal Obstruction

Obstruction of the digestive tract may be congenital or acquired. Atresia – by which is understood a congenital, complete occlusion of the lumen – is by far the most common cause of obstruction in the newborn. The most frequent localizations are the esophagus and the anorectal region. Congenital stenoses and incomplete mucous membranes are much more uncommon. The acquired obstructions dominate in older children.

The children shown here should be easy to identify from the accompanying histories, given in random order:

I. A newborn boy, delivered 6 weeks before term and hospitalized on the suspicion of Down's syndrome. Hydramnios was demonstrated in the pregnancy. After hospitalization there are a few attacks of non-bile-stained vomiting and slight distention of the upper part of the abdomen.

II. A newborn girl, hospitalized for cyanosis and suspected heart disease. During the first 2 days after birth there is no passage of meconium, and the abdomen gradually becomes hard and tense. Two scout films of the abdomen at intervals of 3 hours show increasing amounts of intestinal air but only modest distention and no air-fluid levels. Air cannot be seen with certainty in the colon.

III. A 1-day-old boy, born at term. Since birth there occasionally has been frothy saliva around the mouth and there is plentiful secretion to aspirate from nose and pharynx. Immediately before hospitalization, the patient became grayish in color, with rattling respiration and respiratory retractions.

IV. A boy, born 2 weeks before term and admitted for cyanosis. Clinical examination shows a typical mongoloid infant. The anal opening is missing. The accompanying radiogram was done within the first day of life.

V. A 3-day-old boy, admitted for failure to pass meconium and small attacks of vomiting. After hospitalization there have been several attacks of projectile, bile-stained vomiting. The abdomen is clearly distended, especially in the upper portion. Microscopy of the feces, collected by means of a rectal tube, shows no lanugo hair.

VI. A 17-month-old girl, who for 2 days has had increasing attacks of pain in the abdomen and repeated attacks of vomiting. The temperature is normal.

Compare radiograms and case histories and note how they match. The information is adequate to identify with certainty all patients.

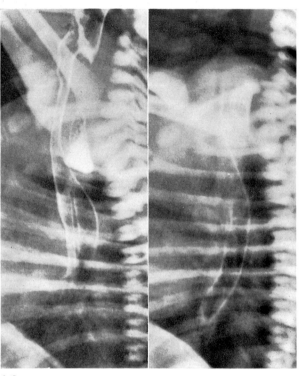

A-2

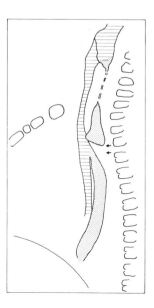

Tracing of the respiratory tract and esophagus on the basis of the two exposures (A-2). The pharynx and trachea are shown hatched, the esophagus is shown gray. The atresia lies between the arrows, while the upper part of the esophagus is contracted.

Answers

A (= III) has *atresia of the esophagus*. The dilated upper esophageal segment is seen as a characteristic translucency in the superior mediastinum. The presence of air in the stomach shows that there must be a *fistula from the trachea to the lower esophageal segment*.

The patient became acutely ill just before hospitalization, probably because of *aspiration*. The upper lobe of the right lung shows the radiographic expression of the aspiration in the form of a triangular consolidation. Aspiration atelectases and secondary pneumonia appear with particular frequency in this region.

Characteristic clinical symptoms in esophageal atresia are excessive and frothy salivation around the mouth and attacks of cyanosis and dyspnea due to laryngotracheal aspiration or regurgitation of gastric contents through the tracheoesophageal fistula. *A-2* shows lateral exposures of another patient with esophageal atresia. Bronchographic contrast material has been injected into the upper esophageal segment through a thin tube. A small amount of the contrast material has been aspirated and is seen passing through the tracheoesophageal fistula with successive filling of the lower esophageal segment.

B (= I) has *duodenal atresia*. Characteristic features are the two partly separated collections of air in the dilated stomach and in the duodenum *(double bubble sign)* and the absence of air in the rest of the intestinal tract.

Operation showed atresia localized to the transition between the first and second portions of the duodenum. Most atresias are localized in the vicinity of the papilla of Vater. The localization of the atresia in relation to the papilla determines whether the vomitus is bile-stained or not.

The suspicion of mongolism was confirmed on chromosome investigation. It is not altogether by coincidence that this infant has duodenal atresia, since more than 25 % of duodenal atresia is found in mongoloid infants. Nor is the presence of hydramnios in the mother a coincidence, as it often is found in cases of gastrointestinal obstruction in the fetus, particularly atresia in the esophagus or duodenum, being due to inadequate resorption of amniotic fluid via the intestinal tract of the fetus in utero.

C (= V) has numerous greatly dilated, bow-shaped loops of small intestine with air-fluid levels distributed throughout the entire abdomen. There is no colonic air. The number of air-filled loops of small intestine and their distribution in the abdomen is a guide to the localization of the obstruction. In this case, it must be an *atresia lying distally in the small intestine*. On operation, it was found distally in the jejunum. It was resected and an end-to-end anastomosis performed. In contrast to atresias of the esophagus, duodenum or anus, which are true developmental defects, most of the atresias of the small and large intestine are considered to be due to ischemic infarction in fetal life.

D (= IV) has an anal or *anorectal atresia*, since according to the case report, no anal opening was present.

The first task, therefore, is to establish the extent of the atresia, as several variants are known. The rectum may be intact, so that only a thin anal membrane is present, or there may be atresia of a greater or larger portion of the anorectal canal. The atresia may be *high,* with the bowel ending at or above the levator ani muscles, or *low,* with the rectum passing in a normal way down through the puborectal sling of the levator ani. In the high atresias, which constitute about 40 % of the cases, it is not unusual to find malformations of the sacrum, sometimes accompanied by neurologic disturbances. Also, other significant malformations are common in these patients. A fistula from the rectum to the urethra in boys or to the vagina in girls is also found most commonly in high atresia. To obtain a radiologic demonstration of the extent of the atresia, the patient may be placed in the Trendelenburg position for a few hours, after which a lateral plain film is taken in the inverted position. This allows air to collect in the dilated colon above the atresia, and if an opaque marker is placed in the anal region, the distance can be measured. However, the results of measurement may be misleading because of inadequate unfolding of the bowel or impaction of meconium above the atresia. The same uncertainty is present in measuring the distance with the aid of ultrasound. The accompanying film was taken after transperineal injection of water-soluble contrast material into the air-filled rectum during fluoroscopy. For radiologic differentiation between high and low atresia, a line sometimes is utilized, drawn from the symphysis to the lower border of the 5th sacral segment. *D-2* shows a film of a patient with membranous imperforate anus, carried out by the technique described above. A small amount of barium has been rubbed into the anal region.

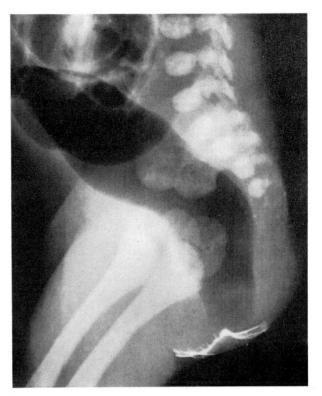

D-2

E (= II) has *meconium plug syndrome.* An enema showed the entire large bowel to be filled out with meconium, which made it difficult to perform the enema until the intestine gradually became dilated. At the end of the examination, plentiful amounts of meconium were passed, following which the patient's condition normalized. A number of patients with meconium plug syndrome later are found to have cystic fibrosis of the pancreas or Hirschsprung's disease.

Barium suspended in physiologic saline may be used for the enema, or diluted water-soluble contrast material.

F (= VI) differs from all the other patients with regard to age, as acquired obstructions now dominate numerically. A radiogram of the colon shows a typical *ileocolic invagination,* with the head of the invagination (the intussusceptum) in the transverse colon. The hydrostatic pressure of the barium enema partly reduced the invagination until only 4-5 cm were lacking *(F-2).* As complete reduction was not possible, the patient was operated on. An ileoileocolic invagination was found, which was easily reduced. The reason for the invagination could not be demonstrated.

Invaginations of less than 24 hours' duration may be reduced in 70-80 % of the cases by the hydrostatic effect of a barium enema. If the symptoms have lasted for more than 24 hours, the frequency of reduction falls to about 50 %.

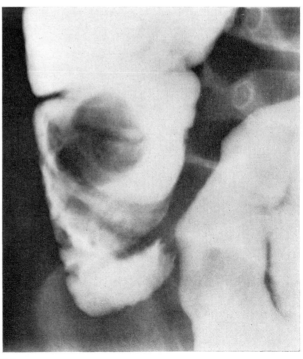

F-2

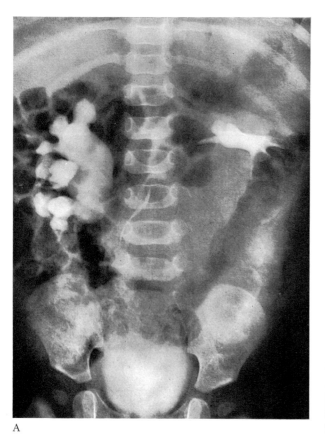

A

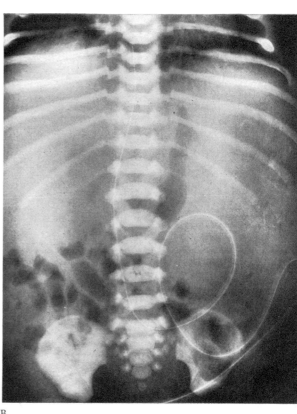

B

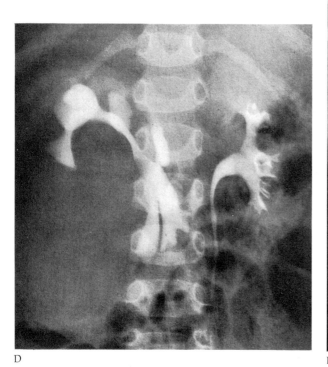

D

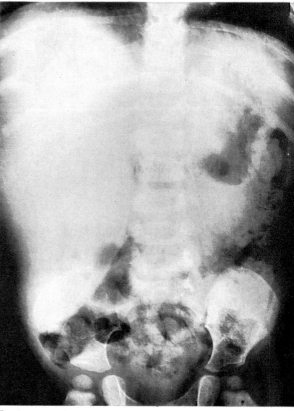

E

Space-Occupying Masses in the Abdomen

The demonstration of space-occupying lesions in the abdomen often presents a considerable number of problems of differential diagnosis. Congenital anomalies, inflammatory changes, benign or malignant tumors of retroperitoneal or intra-abdominal origin are only some of the diagnostic possibilities. Of the radiologic investigations, those most often of use include plain films of the abdomen, intravenous urography, barium studies of the gastrointestinal tract and angiography. Ultrasonic scanning often is of value.

The aim of radiologic examinations is, above all, to establish the anatomic origin of the mass and its nature if possible. In a number of cases, the diagnosis cannot be made until exploratory laparotomy and histologic examination have been done.

A is a 6-month-old girl, hospitalized for fever lasting for 1 week but otherwise with no complaints. Physical examination reveals nothing abnormal, but 5 days later it is considered that a mass can be felt in the position of the left kidney. Urine microscopy is normal.

B is a 2-day-old girl, hospitalized with slight general cyanosis and respiratory distress. A large mass is felt in the left side of the abdomen. The liver is not definitely enlarged.

C is a 1¹/₂-year-old girl, without symptoms until 18 days before hospitalization, when she started having fits of crying, mainly at night. An irregular, lumpy mass is felt in the middle of the abdomen.

D is a 2-year-old girl. During the past 2 months she has been ailing, has had periods with foul-smelling diarrhea and has not increased in weight. Three days before hospitalization, some other children stepped on her stomach and since then she has had pain in the abdomen. On admission, a large mass is found in the right side of the abdomen, about 7 x 14 cm in size. Urine microscopy is normal.

E is a 19-month-old boy, hospitalized for failure to thrive and increasing enlargement of the abdomen. He is not affected, but height and weight are below normal. The abdomen is large, with increased venous markings. Physical examination shows the liver to be enlarged, smooth and sharply edged. There is no demonstrable enlargement of spleen, kidneys or lymph nodes.

F is a 2¹/₂-year-old boy, well until 6 weeks previously. Then increasing tiredness with poor appetite and considerable loss of weight. During the past 5 days he has had intermittent pain in the left side of the abdomen. On physical examination, a lumpy mass is felt in the left side of the abdomen. There is no enlargement of the liver, but possibly of the spleen. Small lymph nodes can be felt in the axillae, inguina and on both sides of the neck.

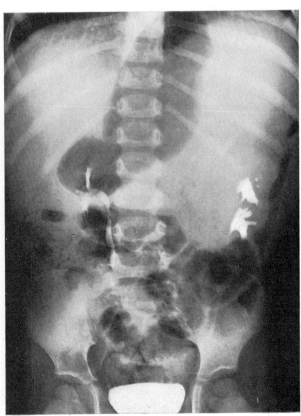

C

F

Answers

A has a strikingly small renal pelvis on the left side, displaced and compressed by a large space-occupying mass in the lower pole of the kidney. The abnormal shadow reaches down in front of the ileum and extends medially in front of the spinal column. Note the contrast-filled segment of the left ureter in front of the 3d and 4th lumbar vertebrae. The right renal pelvis is dilated and the right ureter is not visualized.

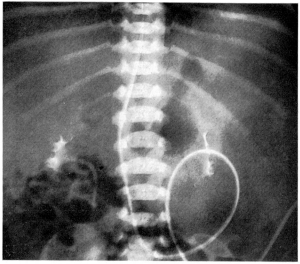

B-2

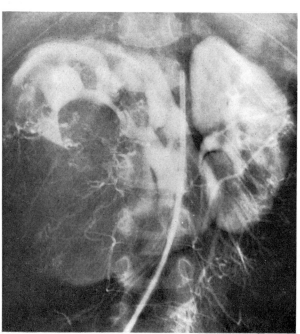

D-2

This could be a left-sided Wilms' tumor, but various circumstances favor another diagnosis. The small size of the left renal pelvis suggests the presence of a duplication. Laterally in the soft tissue shadow below the contrast-filled renal pelvis there can be seen some faint, curved shadows of contrast material that were not found on a plain film of the abdomen. Such crescents sometimes are seen in large *hydronephrosis* or *pyonephrosis*, and ultrasound scanning strengthens this presumption. Operation showed a duplication of the renal pelvis and ureter, with a large pyonephrosis in the lower pelvis. The inflammatory changes extended up into the wall of the upper pelvis, so the whole kidney was removed. A stricture was demonstrated at the origin of the ureter from the hydronephrotic lower pelvis and a normal ureter from the upper pelvis. Four months later, the patient was operated on for *right-sided* ureteropelvic stenosis by means of Y-V-plasty.

A scout film of *B's* abdomen shows a large soft tissue shadow on the left side, displacing the stomach air medially and extending from the diaphragm down to the iliac crest. In the middle of the soft tissue shadow it is just possible to distinguish some patchy *calcifications*. Intravenous urography *(B-2)* shows the left kidney displaced a little medially. Ultrasound indicates the presence of a large mass in the position of the spleen. Laboratory investigations provide no further diagnostic information. An enlarged spleen with small calcifications may be found in venous thrombosis of the spleen, chronic inflammation (particularly tuberculosis and histoplasmosis), intracapsular hemorrhage and processes such as hemangioma and lymphangioma. When the patient's age is taken into consideration, the two last-mentioned possibilities undoubtedly are the most likely, but neuroblastoma, originating from the left adrenal gland, should be included as a differential diagnostic possibility. However, operation showed *hemangioendothelioma*, 9 x 3 x 3 cm in size, *originating from the left lobe of the liver*, but otherwise normal conditions. The tumor was well delimited and was resected. Histologic examination showed scattered calcifications in the tumor.

C's urography shows normal kidneys, but right across the abdomen there are numerous small calcifications. A lateral view of the upper part of the abdomen *(C-2)*, taken before the injection of contrast material, shows the calcifications even more clearly. The most obvious diagnosis is neuroblastoma. Operation showed a tumor that was free of the left kidney but which had grown together with the right kidney. In addition, there were enlarged lymph nodes along the aorta. Biopsy showed neuroblastoma and the patient was given cytostatic chemotherapy but died 7 months later. Neuroblastoma *(sympathicoblastoma)* originates from the medulla of the adrenal gland or from the sympathetic chain. It is particularly frequent in children under the age of 4 years and is the most common malignant neoplasm in the abdomen in this age group. The tumor metastasizes mainly to the liver and bones.

D has been exposed to abdominal trauma of unknown force, so that there is a possibility of a renal lesion with local hemorrhage. Intravenous urography shows a large, well-delimited soft tissue shadow in the position of the right kidney. It displaces and distorts the right renal pelvis, which, in addition, is dilated, suggesting compression of the ureter. This does not resemble the changes seen in rupture with subcapsular hemorrhage, but is more likely to be a *Wilms' tumor*, with solitary cyst as a differential diagnosis. The left kidney is normal. Renal angiography *(D-2)* shows numerous tumor vessels, and subsequent ultrasound scanning confirmed the presence of a large, solid mass in the position of the kidney. At operation, a Wilms' tumor was removed, 8 x 11 x 7 cm in size, apparently radically. The patient was given postoperative cytostatic chemotherapy and high-voltage radiotherapy.

Wilms' embryoma *(nephroblastoma)* is almost as common as neuroblastoma in small children and occurs in particular in children under the age of 5 years. It sometimes occurs bilaterally (about 5 %) and metastasizes almost exclusively to the lungs.

E. A scout film of the abdomen shows a large soft tissue mass with a sharply delimited lower border displacing the air-filled colon. According to the findings on palpation and from the radiologic examination, the differential diagnosis must concentrate on primary tumor of the liver, malignant systemic disease or a storage disease. Both isotope scintigraphy and ultrasound show hepatomegaly without demonstrable focal changes. Comprehensive laboratory investigations suggest glycogenosis (hypoglycemia, hyperlipemia, ketonuria, no rise in blood sugar following the injection of epinephrine and glucagon). Analysis of erythrocytes and leukocytes, followed by liver biopsy, confirmed the presence of *glycogenosis, probably type III.* Glycogenoses are metabolic disturbances caused by enzyme defects and are recessively inherited diseases. Today, several types are recognized. The classic form is known as von Gierke's disease, which corresponds to type I, whereas type II is identical to Pompe's disease. The various types are distinguished, first and foremost, by definite, characteristic glycogenolytic enzyme defects, but, in addition, by differences in course and prognosis.

F's radiogram shows a large space-occupying mass in the left side of the abdomen, displacing the left renal pelvis laterally. Furthermore, the pelvis is slightly dilated and distorted in a way that might arouse suspicion of a duplication. The lateral view *(F-2)* shows no forward displacement of the kidney. The findings probably represent a *retroperitoneal process of extrarenal origin.* Even though there are no demonstrable calcifications in the soft tissue mass, neuroblastoma is a probable diagnosis. On exploratory laparotomy, a very large retroperitoneal tumor was found, which had grown together with the left kidney and extended all the way to the anterior abdominal wall. Freeze microscopy showed *lymphosarcoma,* so high-voltage radiation therapy was started. Lymphosarcoma is more uncommon in children than in adults. On the other hand, it often runs a more fulminant course. In this case, the patient's condition deteriorated so rapidly that effective treatment could not be carried through, and the patient died 2 months later.

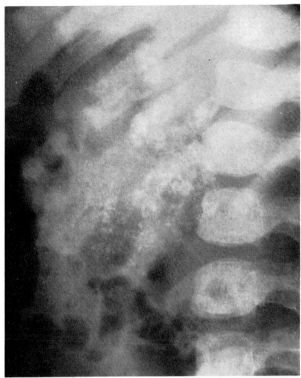

C-2

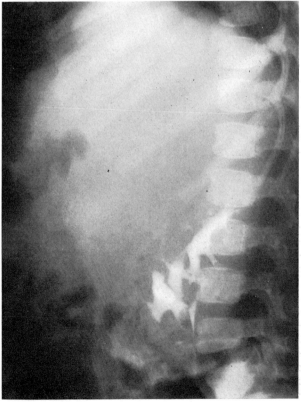

F-2

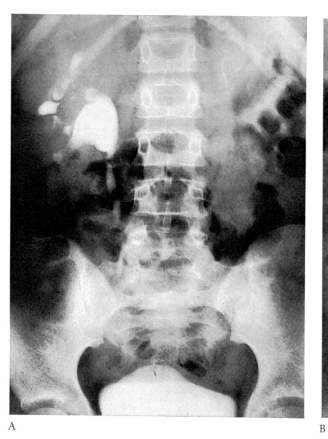

A

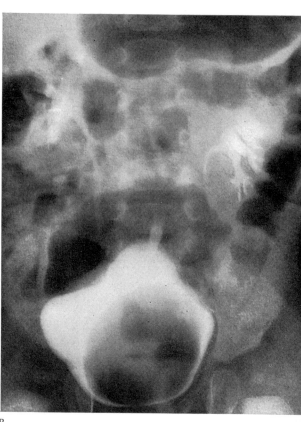

B

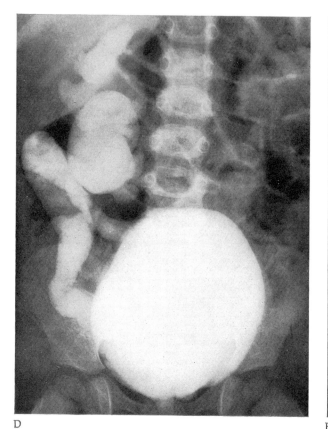

D

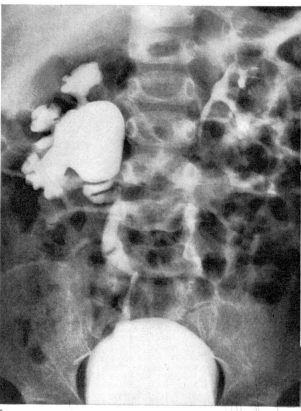

E

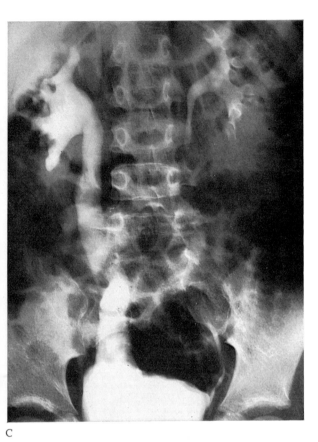

C

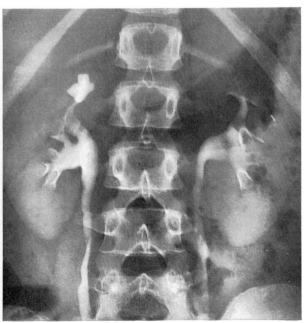

F

Urinary Tract

Several of the patients on these pages present changes in the urinary tract that undoubtedly are due to congenital malformation. In many cases, urographic findings do not present the malformation as such, but secondary changes, usually due to an obstruction to flow, possibly combined with infectious changes and the effect of reflux.

In many older children with recurrent infections of the urinary tract, we are not able to recognize a congenital malformation. However, if the opportunity has not been presented to follow the development of the pathologic changes in the kidneys or in the ureters, it often may be difficult to decide whether a congenital malformation has been present primarily or not.

A is an 11-year-old girl who has been hospitalized 3 times for recurrent pyuria in the course of the past 2 years. At times she has received chemotherapy, antibiotics and mandelic acid granulate. Serum creatinine and blood urea are normal and there is no elevation of blood pressure.

B is a 3-month-old girl. At the age of 6 weeks she had a urinary tract infection that disappeared on treatment with sulfonamides. She now is hospitalized with diarrhea, and clinically she has gastroenteritis, but leukocyturia and significant bacteriuria are demonstrated in addition.

C is a 6-year-old boy, hospitalized for intermittent fever and pyuria over a period of 6 months, treated with sulfonamides with good effect. Significant bacteriuria is found on admission.

D is a not quite 2-year-old boy. As a newborn, he was operated on for anorectal atresia and rectourethral fistula. He is failing to thrive and during the past 6 months has had recurrent urinary tract infections with fever. The accompanying radiogram is from micturition cystourethrography.

E is a 2-year-old boy, hospitalized for acute abdomen and with suspected urinary tract disease. He has had repeated attacks of otitis media but no urinary tract symptoms until 4 days before admission, when the urine was dark and cloudy. No fever, but pain in the right side of the abdomen.

F is a 9-year-old girl, admitted at the age of 5 years for recurrent urinary tract infections over a period of 1/2 year. At that time, intravenous urography was normal, but MCU revealed reflux on both sides. She since has received almost constant prophylactic treatment with chemotherapy up to 1 1/2 years ago. The urine has been sterile since the commencement of treatment and also after terminating the treatment, but, as she still shows reflux, she is admitted for operation.

Answers

A has a *dilated renal pelvis on the right side* with *club-shaped calices.* The renal shadow is rather large and its boundary slightly scarred. On the left side it is just possible to recognize some faint streaks of contrast material at the level of the first lumbar vertebra, but a renal pelvis cannot be outlined. *A-2* is an MCU with heavy *reflux to the left ureter* and filling of a dysplastic renal pelvis with closely packed calices. There is no reflux on the right side. The left kidney is quite small (with a

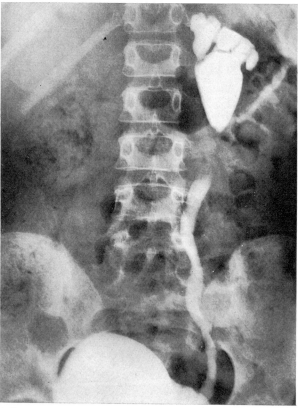

A-2

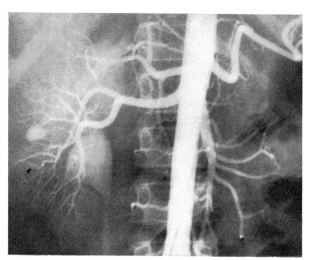

A-3

long axis of 6.5 cm compared to 12.5 cm of the right kidney) and a parenchyma measuring only 3-4 mm. Angiography *(A-3)* shows a thread-like left renal artery branching in the lower pole of the kidney.

These changes did not develop in the course of the 2-year period for which we have information, as urography is available from the first hospitalization and the only difference from this is that the right renal pelvis has become slightly more dilated since then. We may assume that this patient has a slight ureteropelvic obstruction on this side and that a chronic pyelonephritis that first became clinically manifest during the past 2 years probably has contributed to the deformation of the calices and the scarring of the kidney. How the left-sided changes have developed is still more uncertain. A congenital hypoplasia of the renal artery or a thrombosis of the renal artery in infancy is a possible explanation, but, here, infectious changes in the renal parenchyma probably have contributed also.

B. In small children, intestinal air often makes it difficult to obtain a good presentation of the urograms and in particular of the parenchyma. However, *B* has an obviously abnormal displacement of the left renal pelvis downward. At the same time, it is just possible to see a duplication of the right pelvis with a quite small upper pelvis (see tracing *B*). The displacement on the left side could be due to Wilms' tumor in the upper pole, but the presence of a right-sided duplication makes it more probable *that there is also a duplication on the left side,* with a *hydronephrotic, nonfunctioning upper pelvis.* The large translucency in the bladder is a *ureterocele,* which occurs particularly frequently in connection with a duplication of the ureter and especially in relation to the ureter from the upper pelvis, opening abnormally low into the bladder. If the ureterocele is left-sided, this explains satisfactorily the presence of a hydronephrosis of the upper left pelvis. Operation confirmed this supposition. The entire left upper system with associated ureter was removed along with the ureterocele. The patient has been followed up for 2 years and has had significant bacteriuria on only a single occasion. The right kidney and the remaining part of the left kidney have developed nicely. *B-2* shows the conditions 5 months after the operation.

C has a *ureterovesical obstruction.* This has resulted in a rather strong dilatation of the ureter and renal pelvis. MCU showed no reflux, nor was it expected. Operation revealed a pronounced stenosis of the distal end of the right ureter. The stenosed region was removed and the ureter reimplanted in the bladder as in an antireflux operation.

D has a heavy *reflux to both ureters and renal pelves.* This is not a case of duplication of the right pelvis and ureter but a *malposition of the left kidney,* an *ectopia,* so that both kidneys lie on the right side. Note that the ureter from the ectopic left kidney crosses the vertebral column to open normally into the left side of the bladder. *D-2* is from intravenous urography. The opacity of the urograms is less than normal, partly as a result of a

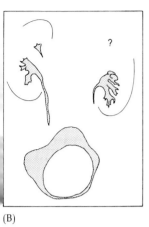

(B)

slightly reduced renal function and partly because of dilution of the contrast material by the great amount of urine present in the renal pelves and ureters. The parenchyma is clearly narrowed and several deep scars are seen. Neurogenic dysfunction of the bladder was demonstrated in addition, with the retention of large amounts of urine, so that a Bricker bladder was constructed, i.e., the ureters were implanted in an isolated segment of the small intestine, with ileostomy, functioning as a bladder.

Now aged 15 the patient still has recurrent urinary tract infections and the serum creatinine is rising slowly.

E's urography shows *hydronephrosis of the right kidney and slight dilatation of the right ureter.* Normal conditions are found on the left side. There is a clear disproportion between the dilatation of the right renal pelvis and ureter, so that ureterovesical stricture or reflux can hardly explain the findings. MCU, in fact, also showed normal conditions. The patient has a congenital *ureteropelvic obstruction.* It might have been expected that the ureter would be narrow or not demonstrated at all, as often is the case. However, it is not altogether unusual to find a moderately dilated ureter with reduced peristalsis, sometimes leading to the erroneous interpretation that a ureterovesical obstruction is present. Such a dilatation of the ureter probably is due to a compromised peristalsis as a result of the presence of the ureteropelvic stricture. In some cases, aberrant renal arteries may be the cause of a bend in the ureter, with hindrance to the flow.

F has a left kidney with completely normal appearance despite the reflux. The right kidney is 2 cm shorter than the left kidney, mainly because of *atrophy of the parenchyma of the upper pole.* In addition, a pronounced *dilatation of the upper calix* is seen.

Such changes may be due to atrophy of the upper renal pole resulting from infection, a congenital stenosis of the central portion of the calix (infundibulum) or to vascular compression of this.

A review of the urography performed 4 years previously demonstrated a slight dilatation of the upper calix and a corresponding narrowing of the parenchyma of the pole. In the present case, therefore, the most probable etiology is a local pyelonephritic lesion of the upper pole.

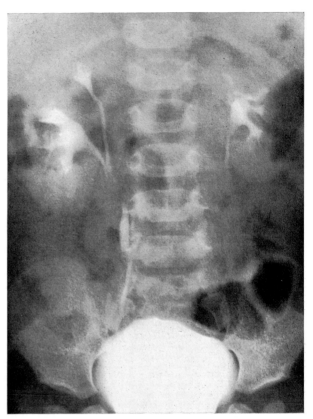

B-2

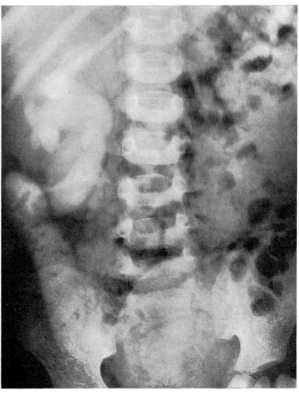

D-2

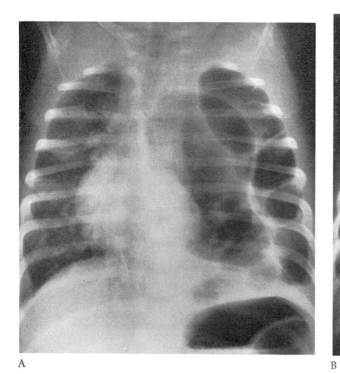

A

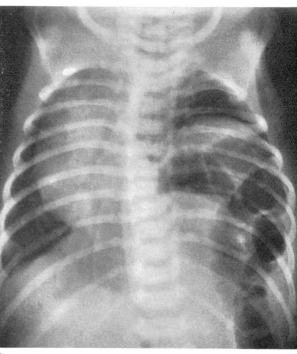

B

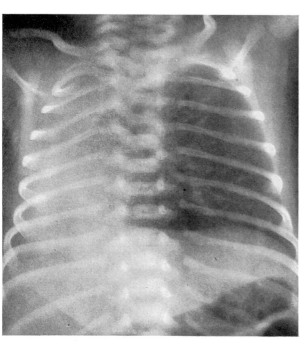

D

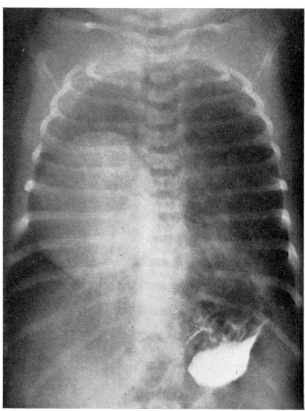

E

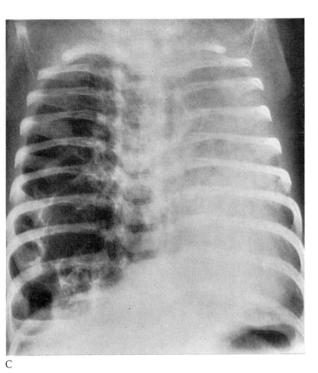

C

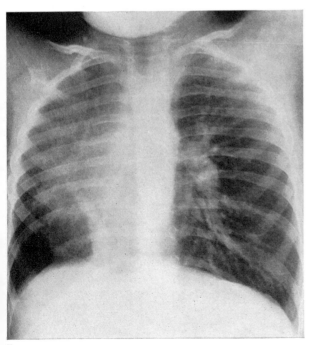

F

Abnormal Distribution of Air in the Lungs

An abnormal increase in the amount of air in a lung or a part of a lung often is due to a check-valvular stenosis of a bronchial branch, allowing air to pass out to the alveoli during inspiration, whereas the passage of air in the opposite direction is more or less obstructed. A check-valve function is not uncommon during infection with plugging of mucus in the bronchial branches, but may also be due to aspiration of foreign bodies, to congenital anomalies or to spasms in the bronchial branches. In all these cases, active dilatation of the lung section is present. Passive dilatation occurs in the form of compensatory emphysema.

The films on these pages do not all fall within the above-mentioned diagnostic possibilities but have been put together to elucidate some of the diagnostic problems.

A is a 6-week-old girl, hospitalized 7 days ago with a rise in temperature and small muscle twitchings in all limbs. Radiologic examination immediately after admission (not shown) revealed small pneumonic consolidations in the left lung. Four days later, several small cyst-like translucencies were found in the lung, and during the next few days she developed increasing respiratory difficulty. The patient received treatment with oxygen and antibiotics. The film shows the conditions on the 7th day.

B is a 2-day-old girl, born at term. She was hospitalized for suspected situs inversus cordis, but there are no clinical symptoms, in particular no respiratory distress.

C is a 17-day-old boy with increasing respiratory difficulty and attacks of coughing lasting for 4 days. Previously he has been well, although he always has had a slightly increased respiratory rate. On admission there is severe tachypnea with subcostal and intercostal retractions. There is no rise in temperature.

D is a newborn girl, born 8 weeks before term. At birth, respiratory distress was present, but she rapidly improved and was placed in an incubator without extra oxygen. Physical examination showed that the heart was lying on the right side. The film was taken on the first day of life.

E is a 1-day-old boy. A few hours after birth there was increasing respiratory distress with cyanosis on crying. Breath sounds were normal over the right lung but the exchange of air over the left lung was strongly reduced.

F is a 9-month-old boy, hospitalized for persistent tachypnea. He becomes breathless on the slightest effort, but not cyanotic.

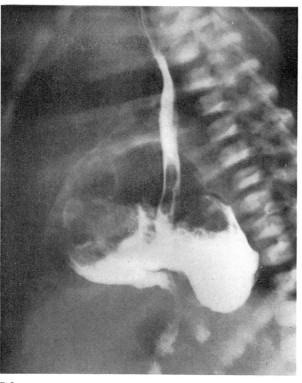

B-2

D-2

Answers

A has several air-filled cavities in the left half of the chest, with displacement of the heart to the right. The changes resemble the picture in diaphragmatic hernia with air-filled intestinal loops in the chest, but the course speaks in favor of *pneumatoceles*. These develop most often in the course of a staphylococcal pneumonia but are also seen in infection with streptococci and *Hemophilus influenzae*. As a rule, pneumatoceles disappear spontaneously in the course of weeks to months. In this patient, the respiratory distress was so pronounced that this was a vital indication for decompression. On tubulation of the left pleura and lung, a quantity of air instantly bubbled out, with immediate improvement in the patient's condition. No further air passed through the drain, which was removed a week later. Radiologic examination 5 weeks later showed normal lungs.

B also has air-filled cavities in the left half of the chest, but delimited above by a diaphragm-like contour. The heart and mediastinum are shifted to the right side. It may be a case of relaxation of the left diaphragm as a result of a lesion of the phrenic nerve, with diaphragmatic hernia as a differential diagnostic possibility. A lateral view following a barium swallow *(B-2)* shows that both the stomach and air-filled intestinal loops are lying in the thorax. At operation, a large defect was found in the posterior portion of the left diaphragm, through which the stomach, large parts of the small and large intestine together with the entire spleen had invaded the thorax – a *Bochdalek hernia*. The abdominal contents were replaced and the defect in the diaphragm closed. After the operation, the left lung expanded nicely.

C has large, thin-walled, cyst-like translucencies in the right lung, displacement of the heart and mediastinum to the left and probably compression atelectasis of the left lung. The history suggests increasing distention of the cysts, presumably as a result of a check-valve effect. The lateral view *(C-2)* shows that the cystic changes occupy most of the right half of the chest. It had been the intention to remove the affected portion of the lung, but the patient's condition worsened, and he died before the operation could be performed. Autopsy showed the changes limited to the right upper lobe, which was converted into large cysts containing several septa. The remains of both lungs, but particularly the lower lobes, were partially compressed. Microscopy showed *multiple cysts of bronchogenic type. A differential diagnostic possibility is cystic adenomatoid malformation*, which is a hamartoma-like malformation in the lung.

D has abnormal expansion of the left lung. The heart and mediastinum are displaced toward the right thoracic wall and the left lung itself extends to the right of the midline. As it is impossible to demonstrate aerated portions of the right lung, *agenesis or aplasia is the most probable diagnosis,* and, in reality, the radiogram is exceedingly characteristic of this. In aplasia, rudimentary remains of bronchial branches are found whereas in agenesis both lung and main bronchus are missing. In the differential diagnosis it is necessary to consider congenital lobar emphysema, complete obstruction of the right main bronchus with atelectasis or a check-valvular obstruction of a large bronchial branch on the left side; for example, because of aspiration of a foreign body. The course of the disease speaks against the last two possibilities, which both give pronounced respiratory distress. For a time, congenital lobar emphysema may run an almost symptom-free course, but subsequently increasing respiratory insufficiency usually develops. After observation for 20 days, during which she remained free from symptoms, *D* was discharged. She managed to thrive satisfactorily and radiologic examination 5 months later *(D-2)* showed conditions practically unchanged. The small triangular shadow laterally in both lung fields is due to a plexiglass screen. Note that there are developmental anomalies of the upper thoracic vertebrae.

E has *congenital lobar emphysema, involving the whole left upper lobe.* The distended lobe displaces the heart and mediastinum to the right and extends retrosternally far over into the right half of the chest. Downward and laterally in the right half of the thorax, aerated portions of the right lung can be seen. The density over the medial portion of the left diaphragm is the compressed left lower lobe.

When 3 days of age, the boy could manage only with 100 % oxygen, so that the thorax was opened and the left upper lobe removed. The bronchus to this lobe was transected at its origin. On inspection of the removed lobe, a pronounced stenosis of this bronchus was found, 4-5 mm from the point of transection. The postoperative course was uncomplicated. In most children with congenital lobar emphysema, the cause of obstruction is not found. As a rule, the emphysema is restricted to one or more segments of one of the upper lobes, most frequently on the left side.

F has pronounced dextroposition of the heart, but we do not gain the impression of a particular hyperaeration of the left lung. The most translucent region actually is the lower portion of the right lung.

A curved, band-like shadow is seen arising from the middle of the right lung, extending down to the cardiophrenic angle. The density is due to a large vein that drains a part of the lung and opens into the inferior vena cava. The appearance is characteristic, being described as the "scimitar sign", and often is observed in connection with *hypoplasia of the right lung.* A hypoplasia can explain the pronounced cardiac displacement. The diagnosis was confirmed on right-sided bronchography *(F-2)*, which shows that the upper lobe and associated bronchial branch are missing. The remaining lobes fill the right half of the chest, so that some of the branches of the middle lobe are directed apically. The partially abnormal drainage of the right pulmonary veins acts as a kind of left-to-right shunt. Subsequent cardiac catheterization showed that the shunt was of modest size, and as the patient's symptoms were in regression, no indications

for surgery were found. Now, at age 15 years, the patient still manages satisfactorily. There is only modest dyspnea on exertion, and never cyanosis.

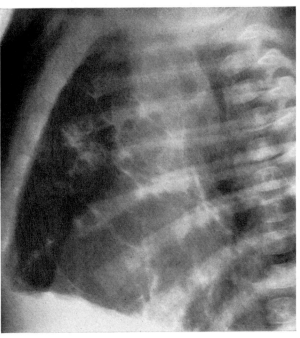

C-2

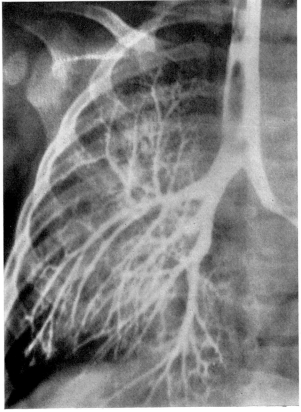

F-2

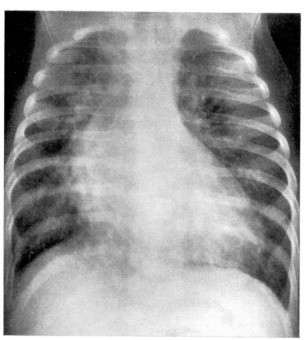

A

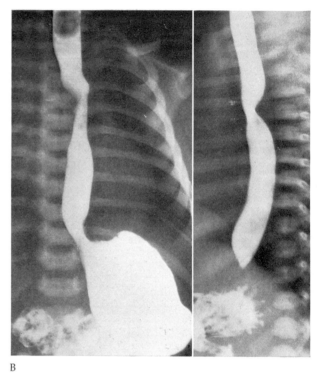

B

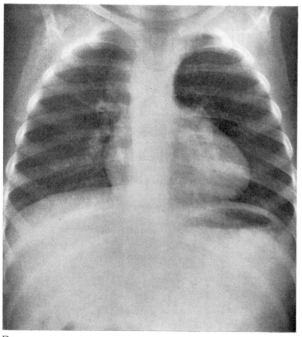

D

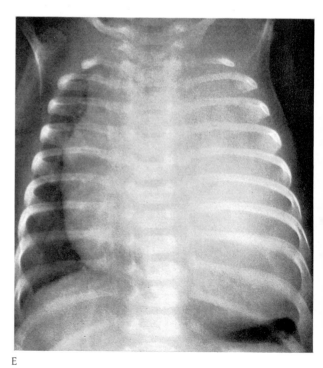

E

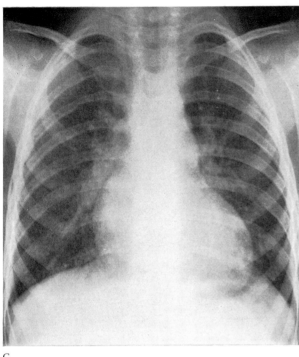

C

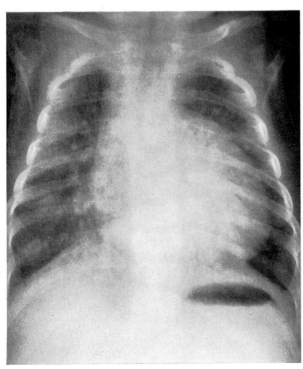

F

Congenital Heart Disease

Congenital heart disease is one of the clinically significant anomalies with the greatest incidence in the newborn. A provisional radiologic diagnosis must be based on an evaluation of the size and shape of the heart and on the appearance of the pulmonary vascularity. Familiarity with the frequency of the various types of cardiac anomalies and the symptoms associated with them, in particular the occurrence of cyanosis, is of importance for the classification. In the first year of life, the size of the heart depends more on the phase of respiration than it does in older children, and a cardiothoracic ratio between 0.39 and 0.65 may be normal.

A is a 9-week-old girl. After a normal delivery, she put on weight satisfactorily and was not cyanotic. Six days before the radiologic examination she developed catarrh, with a rise in temperature and cyanosis. A strong systolic murmur was demonstrated together with tachypnea and general cyanosis but no precordial bulge and no thrill. The ECG showed right axis deviation.

B is a 14-day-old girl. At the age of 5 days, she was admitted with pneumonia, which reacted well to treatment. After this, however, she developed persistent cyanosis, tachypnea and hoarseness, and was unable to give a proper cry.

C is a 2½-year-old boy. At the age of 4 days, he was hospitalized with tachycardia, systolic murmur, a tendency to marbling of the skin and abnormal ECG. Radiographically, the heart was slightly enlarged. At the age of 16 days, he showed signs of cardiac decompensation, which disappeared after digitalization. During the past 2 years he has been thriving well, but the ECG showed development of a combined ventricular hypertrophy.

D is a 15-month-old girl with a systolic murmur but otherwise no cardiac signs. After a few months, a strong holosystolic murmur developed throughout the precordium, most pronounced in the left parasternal line. She also started to be cyanotic, at first in the face but in due course generalized. The ECG showed right axis deviation and right ventricular hypertrophy.

E is a 6-week-old girl. Since birth, she has been unable to cry properly and has had difficulty in drinking. She now has pronounced tachypnea and shows slight reddish cyanosis peripherally. The heart sounds are distant but there is no murmur. The breath sound over the lower part of the left posterior wall is weakened. The ECG is normal.

F is a 4-month-old girl, hospitalized for a recently demonstrated cardiac murmur. She has a pronounced thrill and a harsh, systolic murmur throughout the precordium, maximal in the left parasternal line in the 3d and 4th interspaces. No diastolic murmur and no tachypnea, cyanosis or sign of decompensation. The ECG shows right axis deviation and combined ventricular hypertrophy.

Answers

A's heart is enlarged, egg-shaped, with slightly elevated apex. The superior mediastinum is narrow. The pulmonary vascularity is greatly emphasized. The clinical findings, ECG and radiography all point to *complete transposition of the great vessels.* In this anomaly, the aorta arises from the right ventricle and the pulmonary artery from the left. If the infant is to survive after birth there must be a communication between the two otherwise separate circulations. The late onset of cyanosis in this patient could indicate the presence of a rather large shunt. Angiocardiography with the tip of the catheter placed in the right ventricle *(A-2)* shows the aorta, which arises completely anteriorly and more cranially than normal. A small amount of the contrast material is regurgitated into the right atrium. The main pulmonary artery fills through a large ductus arteriosus (see tracing *A-2).* Cardiac catheterization also suggests the presence

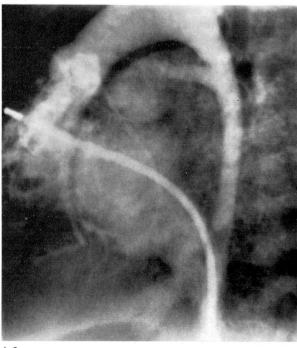

A-2

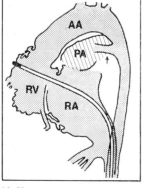

AA: Aortic arch
PA: Pulmonary artery
RV: Right ventricle
RA: Right atrium

(A-2)

of a shunt at both the atrial and ventricular levels. Transposition usually is treated primarily by means of palliative atrial septostomy with balloon catheter by Rashkind's method, followed by a final correction by Mustard's method when the infant has become older. The tendency is toward earlier open heart surgery with primary correction.

B has a *vascular ring,* which causes an impression in the esophagus. Radiologic examination of the thorax showed normal conditions apart from a right aortic arch. Cardiac catheterization was normal. On angiocardiography *(B-2),* the catheter was led via the right atrium and the foramen ovale into the left atrium, where the tip of the catheter was lodged in the auricle. The right-sided aorta is demonstrated and it is seen that the left subclavian artery arises very far distally from the aorta. The impression in the esophagus is caused by this artery. It is possible that in addition there is a ligamentum arteriosum closing the ring around the esophagus and trachea. The patient was operated on at the age of 5 months, and the conditions found were those described. A ligamentum arteriosum passed from the descending aorta immediately in front of the origin of the left subclavian artery and forward to the pulmonary artery, to the left of the esophagus and trachea. The ligament was cut and the left subclavian artery freed as far as possible from the surroundings. She since has been free from symptoms.

C has irregular notching of the inferior border of several ribs, particularly clearly in the right 4th and 5th ribs. The cardiothoracic ratio is 0.55. The increase in width and the shape of the heart suggest left ventricular hypertrophy. The physical examination revealed absence of pulsation in the femoral arteries. *C* has *coarctation of the aorta.* Dilated and tortuous intercostal arteries, functioning as collaterals and transporting blood around the coarctation, produce the erosions of the ribs. The bulge immediately below the aortic knob is due to dilatation of the aorta below the coarctation. Erosions of the ribs rarely are demonstrated in children before the age of 6-8 years. The aortic stenosis usually is localized in the vicinity of the ductus arteriosus. The *preductal type* is most common in infants, and often is combined with hypoplasia of the distal part of the arch (isthmus). In older children and adults, the *postductal coarctation* dominates, often as the sole anomaly. In infancy, any possible decompensation is treated by medication. When the child is fully grown the stenosed segment is resected with end-to-end anastomosis of the aorta. Operation on the preductal coarctation is technically quite difficult, as it often is necessary to use a graft to replace the hypoplastic isthmus.

D's radiogram shows a heart of normal size. The apex is elevated and the waist of the heart accentuated, so that the heart becomes *boot-shaped.* The pulmonary vascularity is reduced. *D* has *Fallot's tetrad.* The aorta is right-sided, as is the case in about 25 % of these patients. The diagnosis was verified subsequently by kine-angiocardiography.

D-2 is a lateral view from the angiocardiography of another patient with Fallot's tetrad. The contrast material has been injected into the right ventricle and passes from here through a stenosed main pulmonary artery. As the aorta overrides a ventricular septal defect, part of the contrast material flows at the same time into the aorta and through the septal defect to the left ventricle. It is also possible to recognize a right ventricular hypertrophy, but this is seen best in the frontal projection, which is not shown here.

The primary surgical treatment consists, as a rule, of establishing an end-to-side anastomosis between the subclavian artery and the pulmonary artery (Blalock-Taussig operation), possibly followed later by a more extensive surgical correction.

E has a large soft tissue shadow at the site of the heart. Angiocardiography was normal. Thoracotomy showed pericarditis with plentiful yellowish serous fluid in the pericardial sac. A tumor the size of a golf ball was found at the origin of the aortic arch. This proved to be a *benign teratoma originating from the pericardium*. The tumor itself cannot be recognized on the chest film. The large shadow is due to the pericardial effusion, probably combined with partial compression of the left lung.

F's heart is enlarged and rounded. The cardiomegaly is due to an enlargement of both ventricles and of the left atrium. The narrow aorta is a characteristic finding. Clearly increased pulmonary vascularity is seen. Cardiac catheterization and selective angiocardiography showed a high *ventricular septal defect* (VSD) with a large shunt from left to right (pulmonary flow 4 times as great as systemic flow). The pressure in the pulmonary artery was only slightly elevated. The patient was treated by banding, an operation in which a tightening band is placed around the main pulmonary artery, so that the pressure falls peripheral to the ligature. When the patient becomes older, a radical operation is performed, with closure of the septal defect. Also in the case of VSD, primary correction now is being carried out at an increasingly early age.

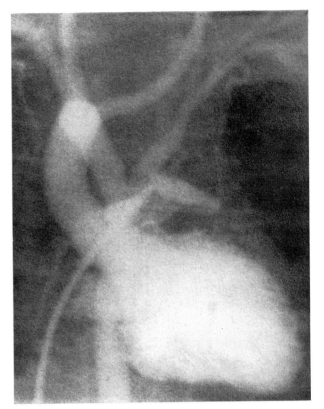

B-2

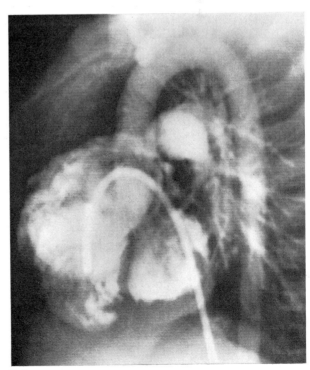

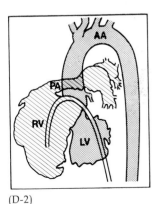

AA: Aortic arch
PA: Pulmonary artery
RV: Right ventricle
LV: Left ventricle

(D-2)

D-2

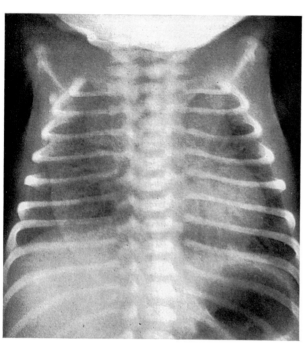

A

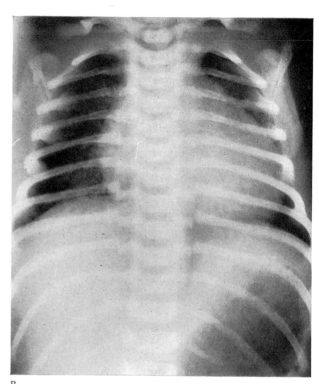

B

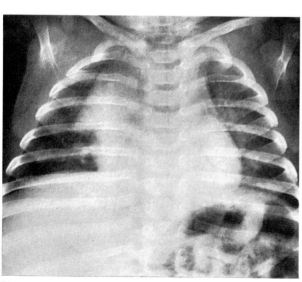

D

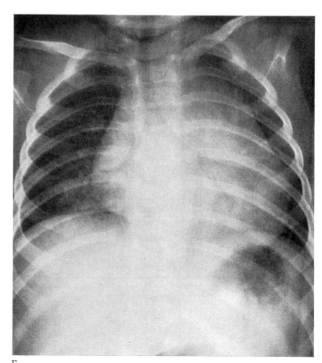

E

Radiologic Diagnostic Errors of Interpretation

Every form of radiologic diagnosis has its pitfalls which trap in particular less practiced but also, at times, more experienced radiologists. Pediatric radiology is no exception. It may be a case of physiologic variations, faulty interpretation as a result of inaccurate centering of the patient or of the x-ray tube or abnormal shadows from objects in or outside the patient. Here, we are not taking into consideration those problems of differential diagnosis that may arise from actual lesions.

The following are a few somewhat more easily recognized pitfalls.

A is a newborn boy, born 4 weeks before term and hospitalized for general cyanosis. Following aspiration of the nose and pharynx, he improves rapidly but nevertheless has to be placed in an incubator with extra oxygen for 6 days. *Are the lungs normal? What is the nature of the shadow in the right lung?*

B is a newborn girl, hospitalized for attacks of cyanosis immediately after birth. On admission, the cyanosis had disappeared, color was nice and respiration normal. *Is the heart normal?*

C is a 4-month-old boy, previously well. For 2 or 3 days he has had a cold, with increasing cough. The temperature on admission was 38° C. *Is the heart normal? – and the lungs?*

D is a 5-month-old boy, hospitalized for pneumonia. For 10 days he has had a cold, with a temperature increase to 38.4° C. Physical examination, including stethoscopy, is normal. *Has he nevertheless pneumonia, or has he aspirated something?*

E is a 7-month-old girl, previously well. Physical examination reveals a slight systolic murmur along the left costal margin. No precordial bulge and no thrill. *The question is whether she has congenital heart disease.*

F is a 7-year-old girl who has had recurrent bronchitis for 2 years. She now is hospitalized with asthmatic bronchitis, a rise in temperature to 39° C and a suspicion of pneumonia. *Are there radiologic signs of pneumonia?*

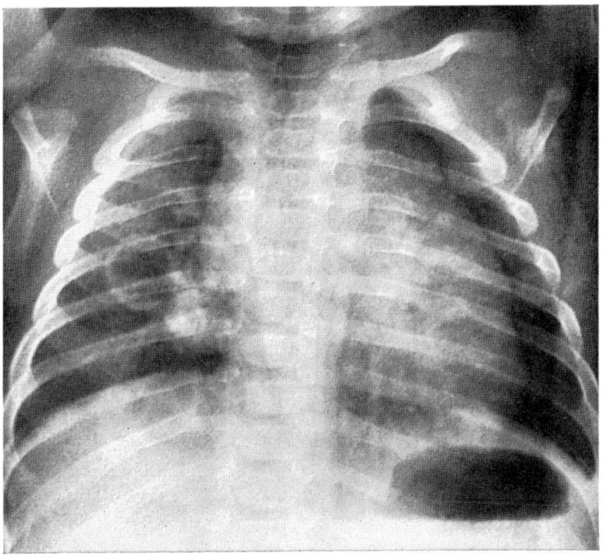

C

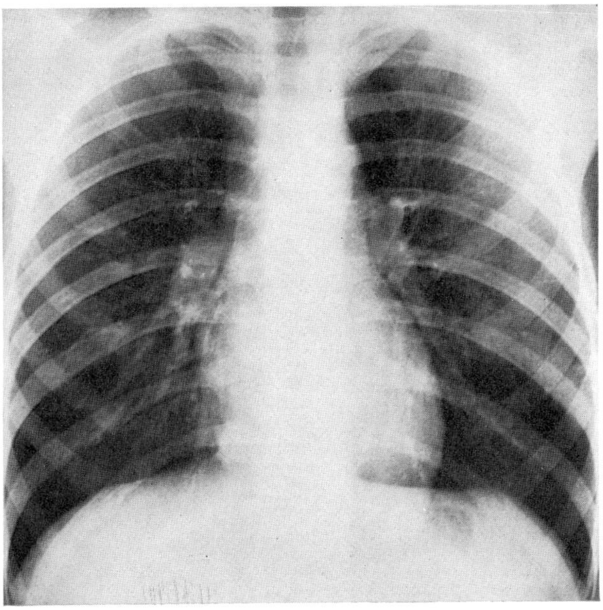

F

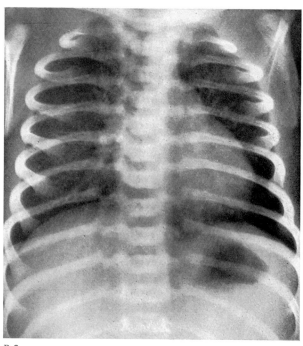

B-2

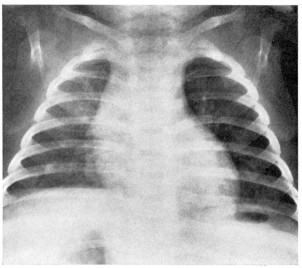

D-2

Answers

A has "wet lungs", with slight linear densities radiating out from both hili, a finding that is fairly frequent in prematures with moderate respiratory distress. The densities represent small atelectases, possibly caused by aspiration of amniotic fluid, and the clinical and radiologic changes usually disappear in the course of 1-2 days. In the lower half of the right lung there is a *shadow with a convex border.* The presence of this shows that the film has been taken with the patient supine (in an incubator), since it is produced by *folding of the plentiful loose skin on the infant's back.* The course of the density is characteristic. Two or three almost parallel folds often are seen on one or both sides. Pneumothorax is the most common error of interpretation.

B has a strikingly large heart with a cardiothoracic ratio of 0.61. Also the supracardiac mediastinal shadow seems to be too wide. However, the film was taken within the *first 24 hours of life.* During this period, the *heart often is particularly large* but usually decreases in size in the course of 24-48 hours. *B-2* shows the same chest 4 days later, where the heart shadow is normal.

C's heart shadow seems rather large, but a characteristic wavy margin of an *enlarged thymus* is seen running down along the left cardiac border, and the cardiothoracic ratio is 0.54 — normal for the patient's age. Just outside the right hilus is a ring shadow and downward in the hilar area there is an intense density. Inflammation? Pneumatocele or other form of cavity?

No, both changes are due to a *pacifier.* The film was taken with the infant suspended in a cylindrical enclosure of plexiglass. While letting out a squall, the infant dropped the pacifier and this now is lying between the wall of the chest and the plexiglass enclosure. If one examines the radiogram of the pacifier *(C-2)*, it is possible to identify the rubber tip as the ring-shaped structure on the chest film, the dense middle piece that is identical to the intense shadow in the hilus and a portion of the large ring of the pacifier, which is just recognizable through the liver shadow (see tracing C).

D has small patchy consolidations perihilar in the left lung. The intense, triangular density outside the right cardiac border is a so-called *sail-shaped thymus*, a characteristic and quite common configuration, which should not be mistaken for atelectasis. *D-2* is the same patient photographed 2 weeks later. The triangular density has become smaller, primarily because the film has been made in deeper inspiration.

E has a strikingly large heart shadow, just as the bronchovascular markings of the lungs appear somewhat increased. *Both diaphragmatic domes,* however, are lying at *a very high level (8th rib),* and the film appears to have been taken at the end of an *expiration,* just as it is probable that the infant has been screaming during the investigation. A repeat film immediately after this *(E-2)* shows the condition in inspiration. The diaphragmatic domes now are lying between the 9th and 10th ribs; the size and shape of the heart as well as the pulmonary vascularity are normal. Otherwise, the infant showed no clinical signs of heart disease, and the slight systolic murmur was physiologic. The two films show the importance of making the radiogram in inspiration.

F has a rather large, almost homogeneous density laterally in the left lung field. Inspection of the patient shows that she is wearing a *ponytail hairdo.* During the investigation, the ponytail has been hanging forward over her left shoulder. *F-2* shows the thorax in the same patient after the ponytail has been moved.

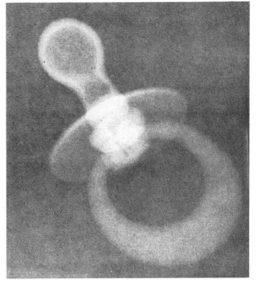

C-2

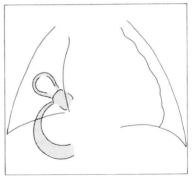

(C)

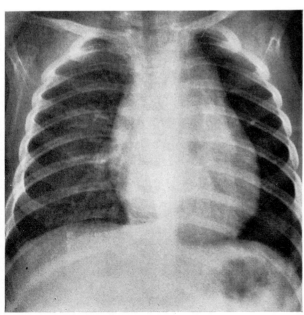

E-2

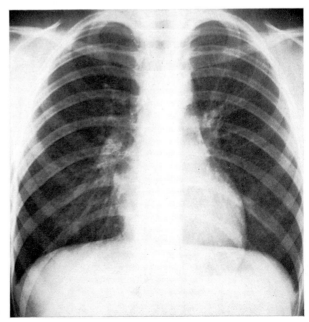

F-2

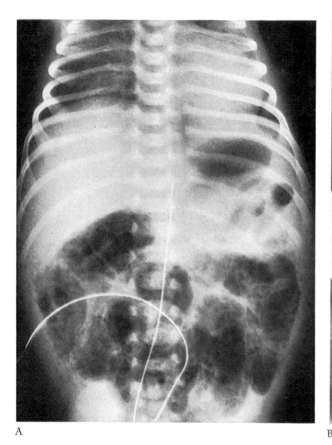

A

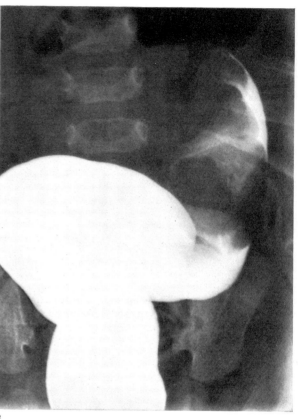

B

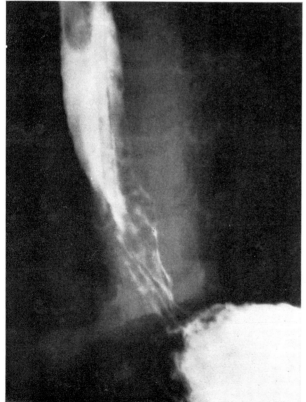

D

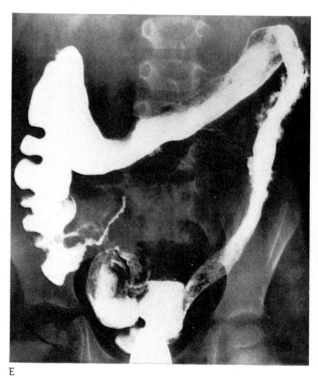

E

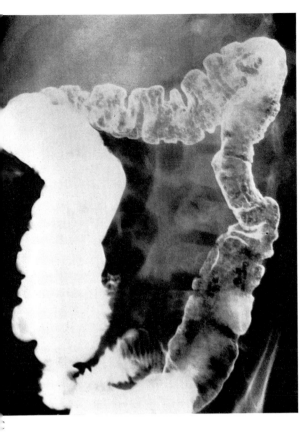

Gastrointestinal Bleeding

Gastrointestinal bleeding occurs in children in all age groups, but there is a rather close relationship between the etiology of the bleeding and definite, more or less well-delimited age groups.

The radiograms on these pages are arranged according to the ages of the patients.

A is a 7-day-old boy, born 10 weeks before term with severe neonatal asphyxia. Immediately after birth, he developed respiratory distress and was placed in an incubator with 85 % oxygen. At the age of 6 days, his abdomen started to become distended and there was fresh blood in the stool.

B is a 5-month-old girl. During the 24 hours before hospitalization she had 4 attacks of thin stool and during the past 5 hours there has been plentiful bleeding per rectum. On admission, she is feeble and pale but not shocked or complaining. Nor were there bouts of crying before admission. A mass is felt in the left iliac fossa.

C is a 6-month-old boy who has been well previously. During the past 6 weeks there have been repeated attacks of thin, offensive stool, black on several occasions, with positive benzidine reaction but without fresh blood. The temperature is normal and the general condition unaffected.

D is an 18-month-old boy with congenital megacolon, demonstrated when he was 1 month of age. The aganglionosis involved the whole descending colon, the sigmoid and the rectum. During the next few months he had recurrent enterocolitis and at the age of 7 months a transversostomy was performed. Postoperatively, he developed a paralytic ileus with pronounced ventricular retention and had to be nourished parenterally. On the 12th postoperative day, hematemesis and melena appeared. His abdomen now is distended and there is pronounced dilatation of subcutaneous veins over the abdomen (caput medusae).

E is a 5-year-old boy who for 3 years has had almost constant watery diarrhea, at times with admixture of blood and mucus. He has been hospitalized on two occasions with medical treatment of his colon disease, but without success.

F is an 11-year-old boy admitted after having observed fresh blood in the stool many times. Occasionally he has had slight diffuse pain in the abdomen, but is otherwise well.

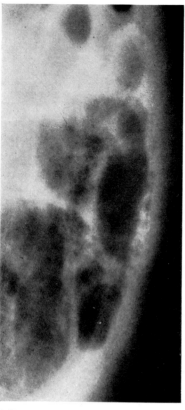

A-2

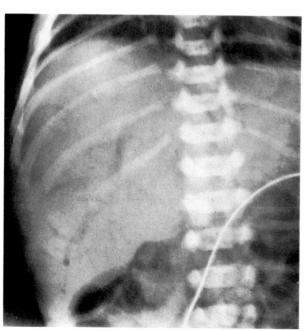

A-3

Answers

A has neonatal *necrotizing enterocolitis*, evidence by numerous small air bubbles along the contour o the dilated air-filled intestinal loops (see also *A-2*).

Necrotizing enterocolitis is seen in premature infants often associated with respiratory distress syndrom (hyaline membranes). The etiology is not certain, but th necrotic lesions, which are localized mainly to the ileun and the commencement of the colon, probably are du to hypoxemia. The lesions usually occur from the 2d t the 4th week of life, but cannot always be demonstrate radiologically. In a few cases, air can be demonstrate in the intrahepatic branches of the portal vein, alone o together with visible air bubbles along the intestinal wall *A-3* shows another premature boy with air in the porta vein branches. The prognosis is poor and seemingly inde pendent of the degree of severity of the radiologic mani festations.

B has *intussusception,* and the head of the invagina tion has reached right down into the sigmoid. In view o the plentiful bleeding, no attempt is made at hydrostati reduction, and operation is performed. An ileoileocoli invagination is found, which is reduced. Numerous large lymph nodes are demonstrated in the root of the mesen tery.

In this patient, the clinical picture was uncharacteristic as there is no report of intermittent abdominal pain, caus ing the infant to cry and draw up his legs. A mechani cal cause of the invagination can be demonstrated in only about 5 %/o of the children operated on; for example, a polyp or hyperplasia of Peyer patches. Invagination occurs almost exclusively in the age group from 3 months to 2 years, somewhat more frequently in boys than in girls

C shows a clinical picture of acute gastroenteritis. The abdominal scout film shows numerous small, uniform round translucencies in the contrast material on the wall of the transverse and descending colon, with a point-shaped central spot of contrast material (see also *C-2*) The changes are characteristic of *benign lymphoid hyper-plasia*. Coloscopy revealed numerous small elevated areas in the sections of the colon in question, together with point-shaped red spots. Radiologic examination 3½ months later showed pronounced regression of the changes whereas, at the same time, the clinical symptoms had disappeared.

Lymphoid hyperplasia is a benign condition that may affect larger or smaller portions of the colon and probably represents the normal reaction of the lymphatic tissue to various stimuli such as infection, allergy and others. Air contrast studies are a prerequisite for a clear presen-tation of the small elevations and, above all, of the char-acteristic umbilication, which makes it possible to dis-tinguish these benign changes from juvenile polyposis and from premalignant familial polyposis (Gardner's syndrome).

D has *portal hypertension with esophageal varices.* The varicosities present as small, oval filling defects in the contrast material in the distal end of the esophagus. Prior to the operation for megacolon, no changes were demonstrated in the liver, but postoperatively the liver function was found affected, with elevated transaminases. The pathogenesis might be damage of the liver parenchyma resulting from or exacerbated by the parenteral nutrition, a thrombosis of the portal vein or an intrahepatic space-occupying lesion obstructing this vein. Percutaneous splenoportography *(D-2)* shows no passage through the portal vein (thrombosis), so that the blood has to be shunted via collaterals to the superior or inferior vena cava. Through some of the collaterals, an increased amount of blood is carried to the veins around the esophagus, which dilates and becomes varicose.

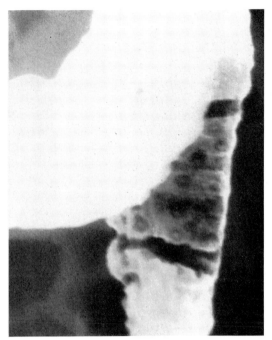

C-2

E has *severe chronic ulcerative colitis,* apparently involving the entire colon but most pronounced at the level of the transverse and descending colon. There are deep ulcerations in the mucous membrane of the transverse colon and the upper part of the descending colon. The smooth, dehaustrated appearance of the intestine suggests advanced inflammation with fibrosis of the wall, changes that gradually convert the colon to a narrow, stiff tube. As medical treatment has failed, operation is decided on. As expected, colitic lesions were found throughout the entire colon, so that subtotal colectomy and temporary ileostomy were performed with a view to subsequently establishing an anastomosis between ileum and rectal stump. Ulcerative colitis is a relatively rare disease in children in the younger age groups but increases in frequency with increasing age. The radiologic findings depend on the duration and degree of severity of the disease, and vary from the demonstration of small, local ulcerations, mainly in the distal end of the colon, to more or less generalized changes with deep ulcerations. At times, the edematous mucosa that remains between the ulcerations may resemble closely packed polyps – *pseudopolyposis.*

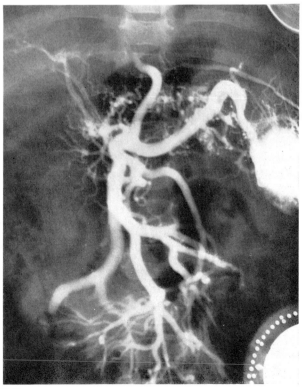

F has two *polyps in the rectum,* one a few cm within the anus, the other about 10 cm higher up. They both are 10–15 mm in diameter. The remainder of the colon was found to be normal. The accompanying radiogram actually should be turned 90°, as it is a lateral recumbent view taken with a horizontal beam, as appears from the two air-fluid levels. Rectoscopy confirmed the presence of the polyps and they were removed 2 days later. On histologic examination, they proved to be *juvenile polyps* without any malignancy.

It should be added that both before and after this admission, polyps were removed from the rectum, on the first occasion when he was 3 years of age and the last when he was 13 years of age. He thus is suffering from juvenile polyposis coli.

D-2

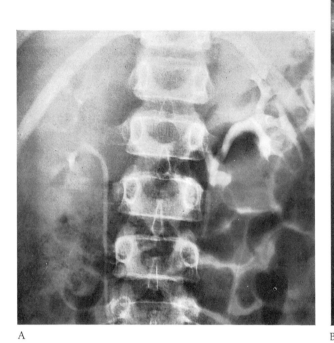

A

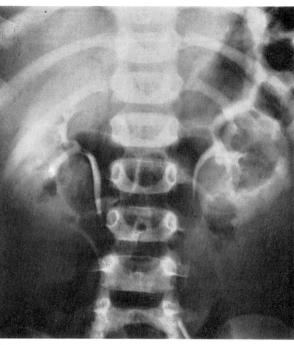

B

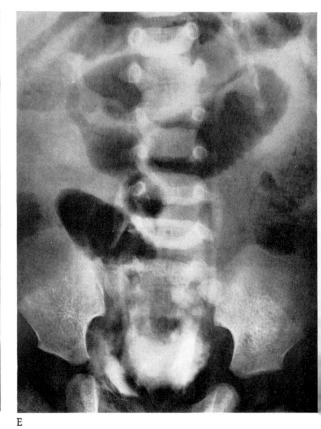

D

E

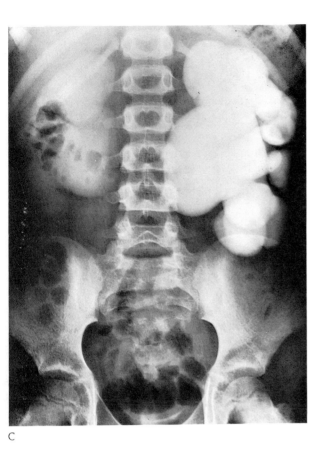

C

F

Blood in the Urine

Hematuria may occur in most diseases of the urinary tract. As a rule, the history, clinical examination and intravenous urography will give the necessary information as to the nature of the hematuria, but if nephroblastoma is suspected, renal angiography and/or ultrasound should be performed. In children, the kidneys are damaged more easily than in adults, since the kidneys are relatively larger and the perirenal fat capsule less developed. Therefore, hematuria is a common finding even in connection with slight trauma.

Here are the case histories in 3 of the children, given at random:

I. An 11-year-old boy has fallen on his stomach from a tree, a height of 3 m. He has pain in the right side of the abdomen and macroscopic hematuria.

II. A 5-year-old boy has been hit in the head with "some lumps of earth" while playing. Since then, he has been sleepy, with amnesia for the event, and has had a few attacks of vomiting. He is hospitalized for concussion. After admission, he experiences pain under the right costal margin, and urine microscopy shows numerous erythrocytes and a few leukocytes. Blood urea is 40 mg/100 ml and serum creatinine 1.1 mg/100 ml.

III. An 11-year-old girl, previously well, has been kicked in the stomach by a horse. She is hospitalized with pronounced tenderness in the right side of the abdomen and in the flank. There is macroscopic hematuria.

In the case of the remaining 3 children, trauma has not been reported.

IV. A 6-year-old boy is hospitalized with macroscopic hematuria and pain in the left side of the abdomen. He had several attacks of urinary tract infection previously, and a left-sided renal duplication was demonstrated with coalescence of the ureteral doubling 3-4 cm above the opening into the bladder. Because of hindrance to flow from the upper pelvis, he had the upper left ureter removed 4 months previously and an anastomosis constructed between the two pelves.

V. An infant aged 2 years and 10 months is hospitalized with an acute abdomen. He is pallid and ailing. For 2 days he has vomited, had attacks of thin stool and refused to eat. The abdomen is tense, distended, with dependent dullness. The urine is bloody. In addition, he has several small blue marks around the right eye.

VI. A boy, not quite 3 years of age, hospitalized for suspected invagination. For 3-4 days he has had thin stool, without visible blood, and occasionally has complained of pain at the umbilicus. A slightly mobile mass, 4-5 cm in size, is felt below the right costal margin. Radiologic examination of the colon is normal. ESR is 7 mm/hour. He has no hematuria.

One of these children has a nephroblastoma, but which one? What is wrong with the other children?

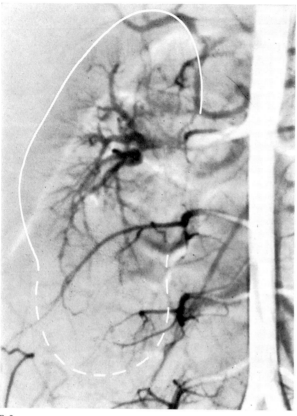

B-2

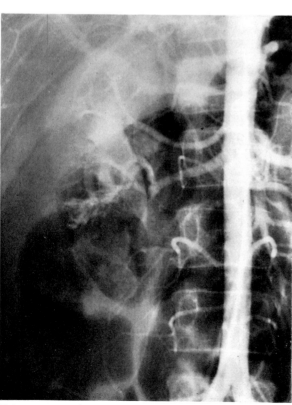

D-2

Answers

A is identical to *I*. Radiologic examination shows that *the right renal pelvis is more slender and not so well filled as the left* whereas the right kidney shadow is normal. Such changes are very common following *slight kidney trauma* and disappear in the course of a few days. In this patient, the hematuria disappeared in the course of 2 days, and control urography on the 6th day showed normal conditions.

B is identical to *VI*. Intravenous urography shows normal urograms on both sides, and the left kidney shadow likewise is normal. The upper pole of the right kidney is well delimited whereas the lower pole cannot be distinguished even by tomography. Ultrasound shows a solid process, 6 x 9 cm in size, which appears to originate from the lower pole of the right kidney. Renal angiography *(B-2)* shows normal vascularization of the upper two-thirds of the right kidney. At the level of the lower pole of the kidney, a few arterial branches can be seen, which appear to be stretched, with very scanty branching, but actual tumor vessels cannot be demonstrated. Together with the results from the ultrasound scanning, the changes are suggestive of *nephroblastoma (Wilms' tumor)*. Operation reveals a nephroblastoma, originating from the lower kidney pole. Since the tumor is very well delimited, it is resected together with the lower kidney pole in macroscopically sound tissue. Postoperatively, high-voltage radiation therapy to the kidney bed and cytostatic chemotherapy are given. After the operation, the right kidney functioned satisfactorily, but 6 months later function is found to be greatly reduced, so that the remainder of the kidney was removed. There was no recurrence, and the impairment in function apparently resulted from radiation treatment. He since has been followed up for 2 years and is in completely good health.

C is identical to *II*. Intravenous urography shows *severe hydronephrosis of both kidneys* and no contrast material in the bladder after 30 minutes. *C* has a bilateral congenital *obstruction at the ureteropelvic junction*. It appears that since he has been quite young he has had episodes of pain in the abdomen and loins several times a year, accompanied by a rise in temperature. During hospitalization, pyuria is demonstrated. Even though *C* may not have been exposed to renal trauma at all, it is not uncommon for a large hydronephrosis or a Wilms' tumor to be found by chance following minor trauma, as such kidneys are more vulnerable to trauma than usual.

C underwent operations in several stages, a new connection being established between pelvis and ureter on both sides. The parenchyma measured only 2-5 mm. He since has been followed up for 7 years. Hydronephrosis still is present, since prolonged dilatation of this size usually is irreversible. However, there no longer is an obstruction of flow and he has not had any urinary tract infection during the past 5 years. Blood pressure is normal. Isotope scintigraphy shows that the function of the right kidney amounts to only about one-fifth of the total function.

In some cases of ureteropelvic obstruction, no organic changes can be demonstrated, and it is considered that the hydronephrosis may be caused by a disturbance of the peristaltic function.

D is identical to *III.* Intravenous urography shows deformation of the lowest part of the right renal pelvis, apparently due to inadequate filling. There is also slight extravasation of contrast material in the same region. The lower pole of the kidney is clearly enlarged, but well delimited. At the upper pole, a double outline can just be seen. There must be a *rupture of the kidney with subcapsular hematoma.* Ultrasound provided information consistent with this. Renal angiography *(D-2)* was performed to establish the extent of the lesions to vessels and parenchyma. A *transverse rupture through the kidney* was found, slightly below the middle, with tear of the artery to the lower pole. Contrast material has seeped out from an arterial branch at the level of the rupture, and the vascularization of the lower kidney fragment is very scanty. Operation confirmed the radiologic findings. The lowest third of the kidney was avascular, lying free, and was removed. The postoperative course was uneventful.

E is identical to *V.* Emergency urography shows slightly blurred but apparently normal urograms, but to the right of the scantily filled bladder *extravasation of contrast material* can be seen. *E* must have a rupture of the bladder or of the posterior urethra. In addition, changes are seen in the upper part of the lumbar spine. *E-2* is a lateral film of this, showing *dislocation between the 12th thoracic and 1st lumbar vertebrae,* with avulsion of the arch and compression of the body of the 1st lumbar vertebra. A fine, linear calcification can just be seen between the body of the 12th thoracic and 2d lumbar vertebrae. This must represent commencing reparative changes and indicates that the fracture cannot be a completely fresh one.

The radiologic findings suggested a further examination of the child's situation and revealed *a battered child syndrome,* while, at the same time, the assaults responsible for the injuries were more or less disclosed. The bladder lesion healed on antibiotic therapy and use of an indwelling catheter whereas a lesion of the medulla resulted in bladder and bowel incontinence.

F is identical to *IV.* A plain film shows three calcifications to the left of the 1st and 2d lumbar vertebrae, suggesting *urinary calculi. F-2* is from the subsequent intravenous urography. Two of the stones are lodged at the anastomosis between the two renal pelves and the third stone is hidden by the contrast material in the dilated upper pelvis. Before the operation there was dilatation of the renal pelves and of the lower ureter, but in the case of the renal pelves the dilatation is moderately accentuated because of spasm around the calculi. The stones were removed, but 18 months later it was necessary to remove two more stones from the same site. The right kidney has been normal throughout all the investigations, and it has not been possible to demonstrate any metabolic anomaly as the etiology of the stone formation.

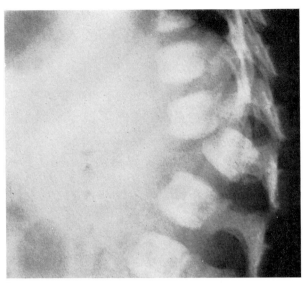

E-2

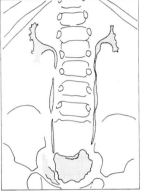

(E)

Tracing of (E). Note the seepage of contrast material to the right of the bladder.

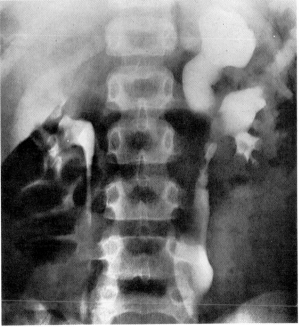

F-2

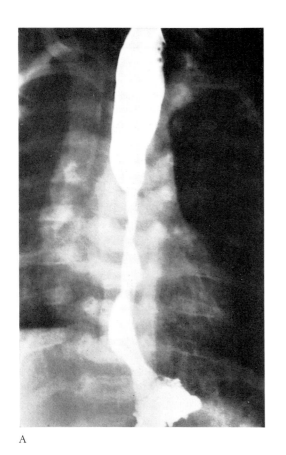

A

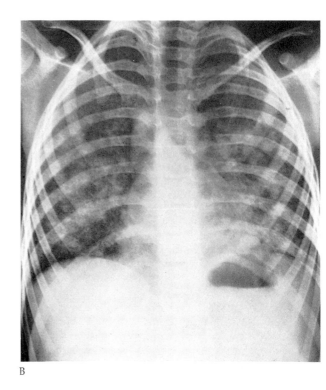

B

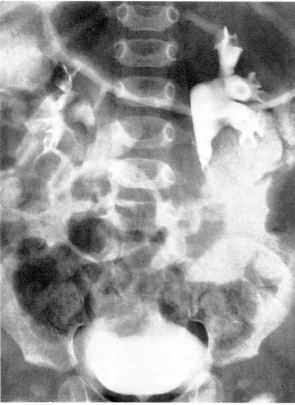

D

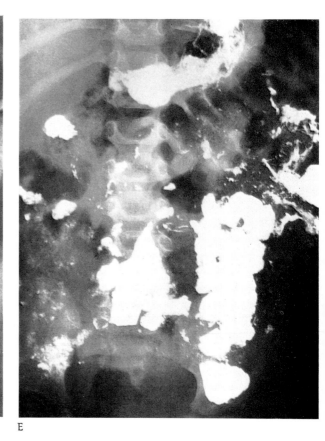

E

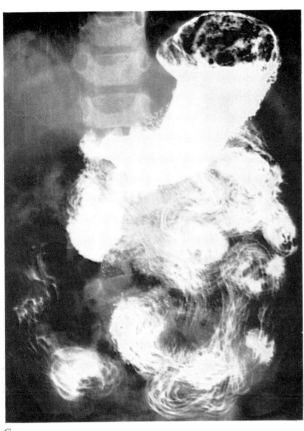

C

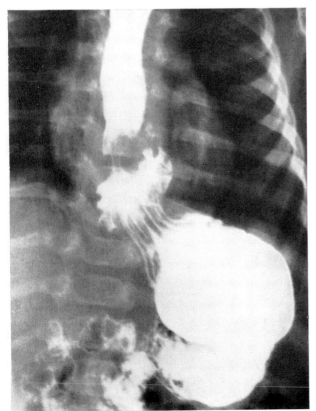

F

Failure to Thrive

Failure to thrive is a common symptom in many chronic diseases that involve inadequate feeding or inadequate utilization of the food.

The 6 children whose films are shown here have only one feature in common, namely varying degrees of failure to thrive. Many of the other children presented in this volume could just as well have been included in the following pages.

A is a 6-week-old boy. Since birth he has had dysphagia, been reluctant to drink and often has had excessive frothy salivation. There have been several attacks of coughing and cyanosis at other than mealtimes.

B is a 4-year-old girl. At the age of 2 years, iron deficiency anemia was demonstrated, which responded well to treatment. Since then there have been several brief periods with coughing and rise in temperature of unknown origin, and she now is hospitalized with fever, cough and vomiting that have lasted for 3-4 days. On admission, she is thin, pale, dyspneic and has slight cyanosis of the lips. Hemolytic anemia and iron deficiency are demonstrated.

C is a 2-year-old boy from Pakistan, who came here 1 month ago. He always has had a poor appetite, and now is hospitalized for failure to thrive, abdominal pain and suspected celiac disease. The stool has been pale but otherwise normal. There has been no vomiting or fever. On admission, he presents as a slender, somewhat undernourished boy. Laboratory investigations show iron deficiency anemia and elevated ESR.

D is a 6-month-old boy, hospitalized for recurrent pyuria and failure to thrive. Pyuria was first demonstrated at the age of 7 weeks, and since then he has been under continuous treatment with chemotherapy. At times there has been severe vomiting and foul-smelling urine but normal temperature. On hospitalization, bacteriuria is demonstrated, so intravenous urography and micturition cystourethrography are performed.

E is a 2-year-old boy. After birth, his development was completely normal, but during the past 6 months it has been characterized to an increasing degree by lack of appetite, failure to gain weight, relative stunting of growth, as well as distention of the abdomen and flattening of buttocks. At the same time, his stool has become frequent (3-5 times daily), thin and pale. The radiogram was taken 1 hour after a barium meal.

F is a 9-week-old boy. One week after birth he began to ruminate and during the past 6-7 weeks there have been increasing attacks of vomiting after meals, this finally becoming projectile. Therefore, he is admitted with suspected hypertrophic pyloric stenosis. He is slender and underweight but not dehydrated. No ventricular peristalsis is observed and no pyloric tumor can be felt. Examination of the stool shows positive benzidine reaction.

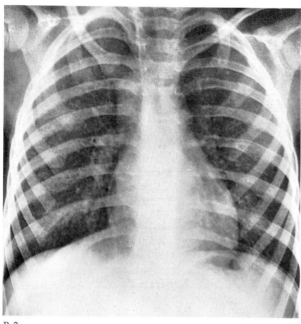

B-2

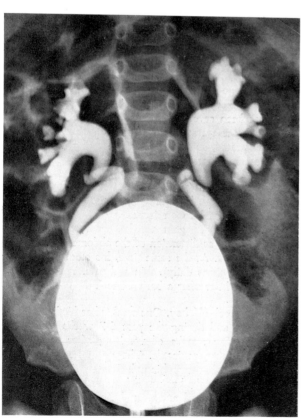

D-2

Answers

A has a 2-cm-long *stricture of the esophagus,* approximately at midlevel. Above the stricture the esophagus is slightly dilated. Although most esophageal strictures develop as a result of reflux esophagitis usually because of hiatal hernia, the present stricture probably is *congenital.* On the one hand, the clinical symptoms have been present from birth, and on the other hand there is only a rather short stenosis, localized a little below the point where esophageal atresia most commonly is found. The investigation did not reveal hiatal hernia or cardioesophageal incompetence.

B has *idiopathic pulmonary hemosiderosis.* Both lungs show coarse, confluent, patchy consolidations, simulating bronchopneumonia. These disappeared in the course of 6 days but left a diffusely increased interstitial pattern, together with scattered, small, granular densities (*B-2*). During the following 18 months she has been hospitalized a further 4 times with cough, suspected pneumonia and anemia. On each occasion, both lungs showed a mixture of large, patchy densities and small, diffusely distributed granular shadows. The diagnosis was first verified at thoracotomy with lung biopsy. In idiopathic pulmonary hemosiderosis, which is relatively rare and mainly seen in children, there are recurrent small hemorrhages into the pulmonary tissues, producing the pneumonia-like consolidations. Released hemosiderin is gradually deposited in the reticuloendothelial cells of the lungs in such amounts that they appear as sharply defined, small, granular shadows, as are also found in pulmonary hemosiderosis in adults with decompensated mitral stenosis. The clinical characteristics of the disease are nonproductive cough, fatigue, pallor and insufficient increase in weight. The sputum sometimes is blood-tinged and may contain heart-failure cells. There is hypochromic anemia, occasionally occult blood in the stool and possibly lymphadenopathy. The roentgen changes may, above all, resemble those in miliary tuberculosis, silicosis and sarcoidosis.

C underwent a radiologic examination of stomach and small intestine because of abdominal pain, and the films shown are from this examination. The small intestine was found to be impacted with *ascarides* (roundworm), presenting as longitudinal filling defects in the barium shadow. Subsequent examination of feces showed myriads of eggs of *Ascaris lumbricoides.* On treatment with piperazine, large amounts of worms were passed, so that for several days the stool consisted almost exclusively of these. An examination of the remainder of the family revealed ascariasis in a further 2 of 4 family members.

D has dilatation of the left renal pelvis whereas the right pelvis and the outlines of both kidneys are normal. On the present radiogram, a primary obstruction of the left ureteropelvic junction might be suspected, but information from subsequent examinations spoke against this etiology. In view of the persistent pyuria, urography was supplemented by micturition cystourethrography (D-2), which shows heavy *reflux to both ureters and pelves*. When the ureters dilate because of reflux or an obstruction to flow, they usually become, at the same time, elongated and tortuous, and a bend at the junction between the left ureter and pelvis probably is the reason for the inhibited flow. The antibacterial treatment was continued with chemotherapy and antibiotics. Repeated examinations during the following years showed decreasing reflux, beginning on the right side. Ten years after the present investigation, radiologic examination showed 2 normal-sized, smoothly outlined kidneys with well-shaped renal pelves, and there was no vesicoureteral reflux.

E has *celiac disease (gluten intolerance)*. A biopsy from the small intestine taken at the ligament of Treitz showed a totally avillous mucosal membrane with a very low content of lactase, sucrase, maltase and alkaline phosphatase. A radiogram (1 hour after a barium meal) shows a pronounced *segmentation of the barium column* in the small intestine, mainly the ileum. The appearance is clearly pathologic for the patient's age group. *E-2* is a film taken 15 minutes after the barium meal. At this moment, the contrast material is mainly distributed in the duodenum and jejunum, where the *mucosal folds are characteristically coarse and thickened*, most clearly seen in the duodenum. After a gluten-free diet for a year there was complete clinical recovery, just as the biopsy from the small intestine became normal both morphologically and enzymatically. A subsequent provocation with ordinary gluten-containing diet for a period of 2 months resulted in pale, bulky stool a couple of times daily, as well as in reduced appetite. The biopsy from the small intestine once again showed flattening of the mucosal membrane, with a low content of enzyme, and thus confirmed the diagnosis of celiac disease.

F has a small *hiatal hernia* of the sliding hernia type whereas the pyloric canal and stomach emptying time are normal. By upright positioning in an infant seat for 1 hour after all meals, and increasing the viscosity of the diet by adding pulverized carob seeds, the vomiting and regurgitation decreased and the patient could be discharged from the hospital for home treatment.

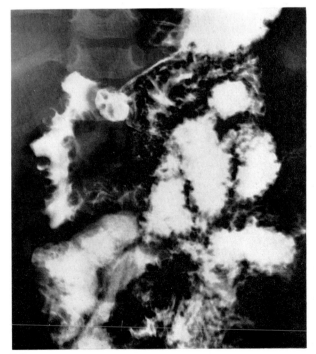

E-2

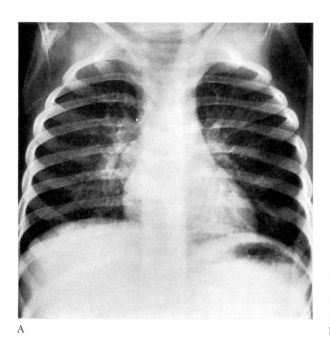

A

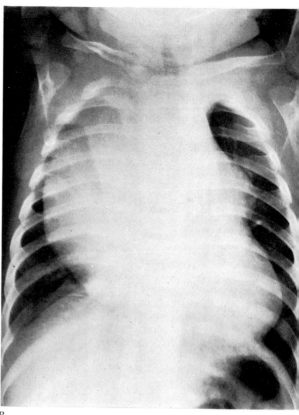

B

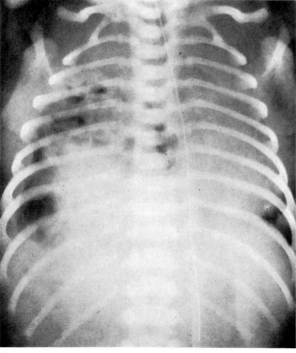

D

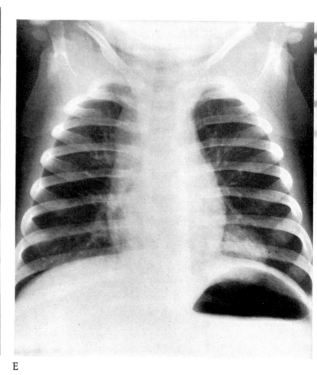

E

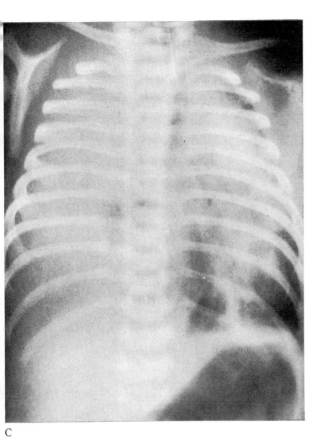

C

Problems of Differential Diagnosis

A is a 16-month-old girl, hospitalized with upper respiratory infection. On admission, no abnormal chest sounds are found. Respiration is free and temperature normal.

B is a 6-month-old girl, hospitalized for suspected pneumonia and asthmatic bronchitis. The last week before admission she had a rise in temperature and slightly labored respiration, treated with penicillin.

C is a 21-day-old girl. She had been well until the day before hospitalization, when she started to be short of breath, developed slight cyanosis of the lips and refused to eat. There was no rise in temperature. As there was respiratory failure on admission, she was treated with tracheal intubation and mechanical ventilation.

D is a newborn boy with respiratory distress and peripheral cyanosis. On physical examination, the heart is found displaced to the left and breath sounds are diminished on both sides.

E is an 8-month-old boy who has had a cold for 3 days with coughing and rise in temperature to 39-40° C. Immediately before admission, he had an attack of convulsions lasting for 5 minutes, with cyanosis and foam around the mouth.

F is a 2-year-old girl who developed a violent attack of coughing while eating hazelnuts 3 hours before the radiologic examination. On hospitalization, she still is coughing and there is some inspiratory stridor and rapid respiration. Diminished breath sounds are heard on the right side.

What are the diagnoses and/or differential diagnoses?

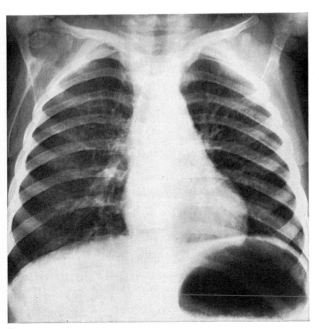

F

Answers

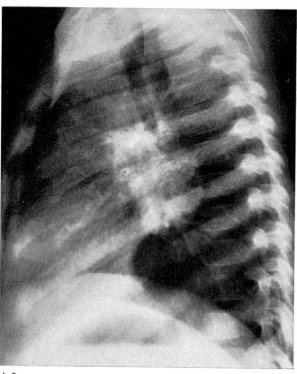

A-2

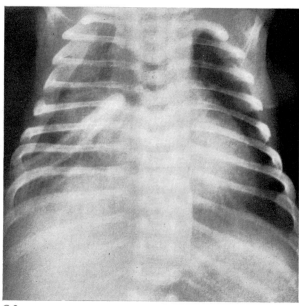

C-2

A has a smooth and well-delimited shadow in the right hilus. It is lying just below the right main bronchus, which is displaced slightly laterad. On a lateral film *(A-2)*, the density is localized downward and backward in the hilar shadow. It could be an enlarged gland, even though further clinical investigations gave no support for this supposition. Other differential diagnostic possibilities are bronchogenic cyst or esophageal duplication. Another examination 2 months later showed the mass quite unchanged. At the same time, it was shown to produce a smooth impression in the esophagus. Thoracotomy revealed a *fluid-filled bronchogenic cyst* 3 cm in diameter.

B has a large intrathoracic mass, which hides most of the heart and displaces it slightly to the left. A lateral film showed that the mass was localized retrosternally. The ECG was normal and angiocardiography showed the heart displaced but otherwise normal. Tumors in the anterior mediastinum almost exclusively are thymoma, dermoid cyst or teratoma. On thoracotomy, a cyst-like tumor was removed. Histologic examination showed it to be a *benign thymoma.*

C's chest film shows that the entire right hemithorax is opaque, with strong displacement of the heart and mediastinum to the left and partial compression of the left lung. As empyema was suspected, a rubber tube was placed in the right pleural space and 250 ml of pleural fluid removed, which was found to be fatty chyle without bacteria *(C-2).* After this, respiration was quickly restored. *C* has *spontaneous neonatal chylothorax* of unknown etiology. After 14 days, the production of chyle became considerably reduced, and the tube could be removed on the 25th day. The patient was discharged from the hospital 1 week later. At this time, only an insignificant pleural thickening remained at the right lateral thoracic wall.

Chylothorax is an uncommon but serious condition. The neonatal form possibly is due to congenital malformation of the intrathoracic lymphatic system. Other possible causes suggested for the condition are trauma, thoracotomy, compression of the thoracic duct and thrombosis of the vein (usually the left subclavian) into which the thoracic duct opens.

D has air-filled intestinal loops in the right side of the thorax, some of which apparently are lying above the diaphragm, as the left diaphragmatic leaf is seen at the level of the 9th rib. The boundary of the right diaphragmatic leaf cannot be identified, and paralysis of the phrenic nerve (birth injury) – with elevation of the diaphragm – cannot be excluded, although it does not explain the presence of intestinal air. In Chilaiditi's syndrome, air-filled portions of the colon are seen interposed between diaphragm and liver, but this occurs particularly in meteorism, and in any case the air in this patient has hardly reached as far as the colon. There remains the possibility of a *diaphragmatic hernia*. Operation revealed a *right-sided Bochdalek hernia*. No hernial sac was found, since the herniation usually occurs through a persistent embryonal canal between the peritoneal and pleural spaces. The right liver lobe and most of the intestine lay in the thorax. They were replaced and the diaphragmatic defect closed. In order to achieve slow expansion of the atelectatic right lung, a pleural tube was not inserted. Despite the large right pneumothorax (D-2) there were no respiratory problems following the operation. The patient was placed in an incubator with 45 % oxygen, and 6 days later he managed well without oxygen.

A right Bochdalek hernia is much more uncommon than a left, probably because the defect in the diaphragm may be closed by the liver. If the defect is large enough, however, part of the liver may protrude through the diaphragm, in rare cases — as in the present patient — together with loops of intestine.

E has had attacks of febrile convulsions. On admission there was upper respiratory infection, rapid and gasping respiration and high fever. Crepitation and slight rales can be heard on the lower left posterior wall of the chest. The radiogram shows a sharply defined round shadow in the left lower lobe. The outline may be followed down through the stomach gas. The opacity is produced by a *segmental alveolar pneumonia*. On treatment with penicillin, the patient recovered clinically in 3 days, and a further radiologic examination on the 4th day showed that the consolidation was fading out. Two weeks later, the lungs had become normal. Round pneumonic consolidations with this appearance are not altogether uncommon in children, and if the history is not known they may give rise to a suspicion of primary or secondary tumor, abscess or fluid-filled cyst.

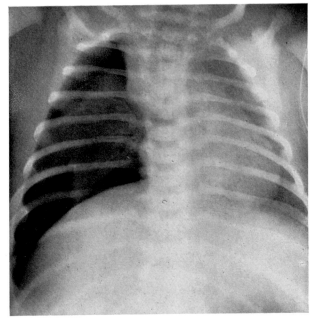

D-2

F has emphysema of the lower portion of the right lung. The lung markings are clearly reduced and the diaphragm flattened. In view of the history, it is probable that a *foreign body* has become lodged in the associated bronchial branch and caused a *check-valvular obstruction*. On bronchoscopy, half a hazelnut is found in the right bronchus, immediately below the origin of the upper lobe bronchial branch. It was removed, and no further foreign bodies could be found, but there was plentiful secretion. Nevertheless, the next morning she coughed up another small piece of nut, after which she recovered.

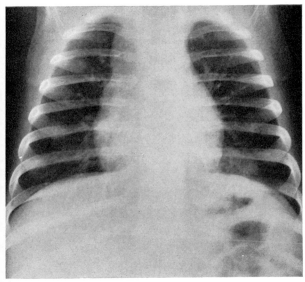

E-2

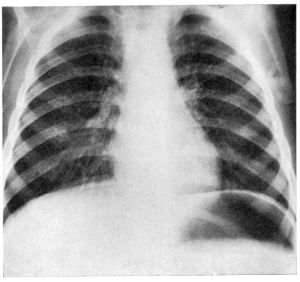

F-2

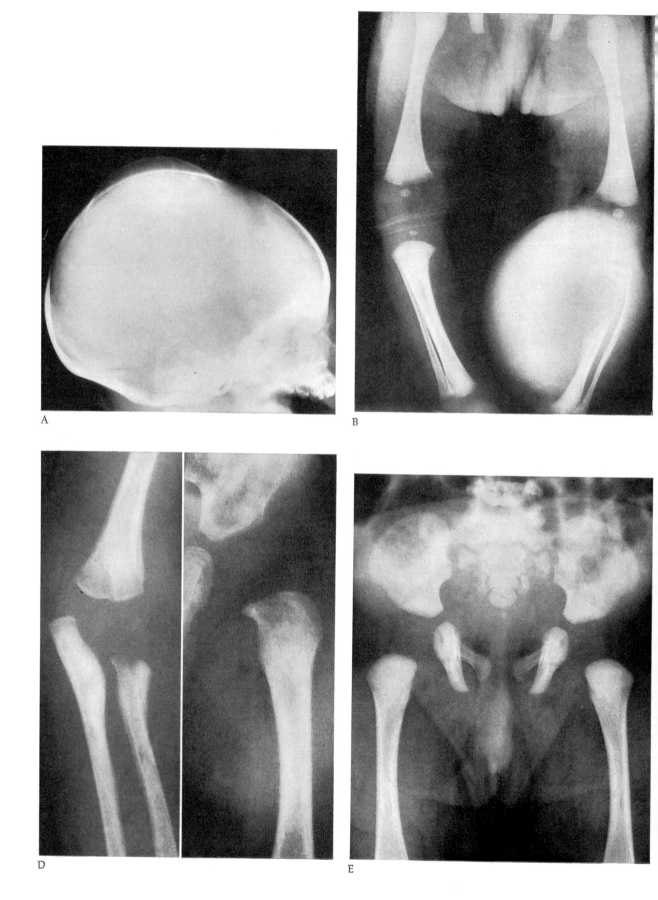

A

B

D

E

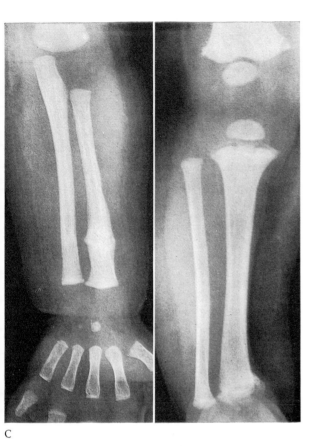

C

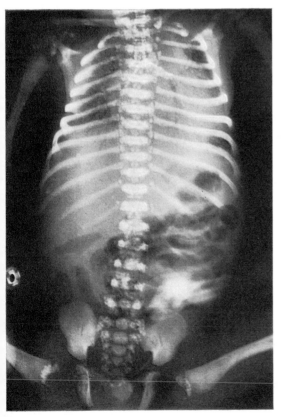

F

The Skeleton

Radiologic changes in the skeleton of the child constitute a very comprehensive topic. Just as the other organ systems, the skeleton is the target of a long series of acquired disorders that are more or less specific for childhood and often for definite age groups. In addition to these there are constitutional (intrinsic) diseases of bones, localized to a single or to a few bones, unilateral or bilateral, often associated with chromosomal aberrations or resulting from induced errors in embryonic development. Finally, there are a large number of generalized dysplasias, which may be of metabolic, nutritive or enzymatic nature, but in the majority of cases are of unknown etiology.

In this section, a few characteristic cases will be shown in which the skeleton is affected. Two are of a traumatic nature, one is infectious, one is neoplastic, one is a dysostosis associated with manifestations in several other systems of the body and chromosomal anomaly and one is a dysplasia of unknown etiology.

The radiologic findings are so varied and individually so typical that further information is not necessary in order to classify the patients within the disease groups mentioned and to make a radiologic diagnosis.

Here are the data for the present patients:
A, a 1-month-old girl.
B, a 1-day-old girl.
C, a 3½-month-old boy.
D, a 1-month-old boy.
E, a 3-day-old girl.
F, a newborn boy.

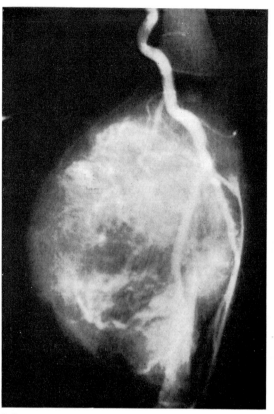

B-2

Answers

A has a smooth bone shadow, projecting beyond the outer table in the parietal region. The changes are due to a subperiosteal hemorrhage – a *cephalhematoma* – caused by trauma to the fetal head during labor. Cephalhematoma is most common in the parietal or occipital region, and in the latter position may be mistaken for meningocele. The hematoma never extends beyond the sutures. In the course of 1-2 weeks after birth, a ring of bone is formed along the periphery of the hematoma, and from here a shell of bone gradually extends along the whole elevated periosteum. Pressure of the hematoma at times may result in localized rarefaction of the underlying bone. The radiologic findings usually disappear completely in the course of a few months, but sometimes leave structural changes that persist for several years.

B was born with a tumor the size of a tennis ball on the lower left leg. The skin is red, tense and with strongly increased venous markings. The radiogram shows pronounced destruction in the proximal end of the tibia, and a highly vascularized tumor is demonstrated on femoral arteriography *(B-2)*. On biopsy, it appears to be a *fibrosarcoma*, for which reason the left leg was amputated by disarticulation in the hip. Since then, she has put on weight satisfactorily, so far for 1 year, has been fitted with a prosthesis and has started to stand on it.

Fibrosarcoma is a rather rare tumor. The most important malignant bone tumors of childhood are Ewing's sarcoma, reticulum cell sarcoma and osteogenic sarcoma.

C has an old *fracture of the left radius* as well as *avulsion fractures of the metaphyses of the distal end of the right femur and of the proximal and distal end of the right tibia,* all relatively new. A skeletal survey showed recent *fracture of the left humerus (C-2)* and *healed fracture of the left 5th rib (C-3)*. In addition, the boy presents a wretched appearance with petechiae and ecchymoses scattered over the head, neck, trunk and limbs. The radiologic findings are characteristic of a *battered child syndrome,* and it proved possible in due course to elucidate, in part, the nature of the physical abuse.

The battered child syndrome is characterized by the presence of skeletal lesions in the form of fractures, dislocations or epiphysiolyses of different ages in children usually under the age of 4 years and most often under the age of 1 year. The most common lesion is avulsion of a metaphysis, resulting from a pull on a limb. Ecchymosis occurs but is not a constant finding. Subdural hematoma is a serious and not uncommon lesion.

D has staphylococcal sepsis with *multiple osteomyelitic foci.* On admission at 14 days of age, local swelling, erythema and tenderness were found over the left elbow, left hip and right thoracic wall. Radiologic examination showed no bone lesions. Blood culture gave growth of staphylococci, so that antibiotic treatment was started immediately. Bone destruction in the left elbow was first observed 1 week later. It appeared as a small area of

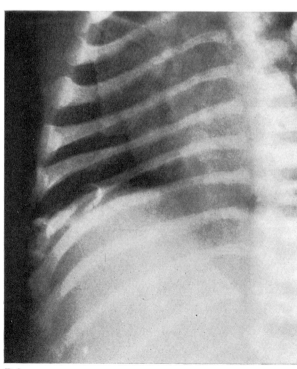

D-2

osteolysis in the proximal end of the radius and a fine periosteal reaction. The radiograms of the left elbow, left hip and right half of the chest (D-2) were taken when the patient was about 4 weeks of age. In the left elbow, the infection has spread to the joint and there are changes in all the metaphyses, both in the form of destruction and in the form of reparative processes. Osteolysis of the hip metaphysis is seen, but the contour of the bone has been preserved. There is pronounced destruction in the right 7th rib, with strong formation of new bone. Two weeks later, the clinical symptoms were subsiding satisfactorily.

If an optimal presentation of the bones under suspicion is available, osteomyelitic lesions in the form of small metaphyseal destructions often can be recognized within a week after the onset of the clinical symptoms.

E has *mongolism* (Down's syndrome, trisomy 21 syndrome). The radiogram shows characteristic changes in the pelvis, as the acetabular roofs are tilted to a more horizontal position and the ilia are larger and flare more laterad than in the normal child. The evaluation is made most easily by measuring the acetabular and iliac angles and comparing these with the normal values for healthy children in the same age group.

The skeleton of the hand shows both retarded bone maturation and often a shortened 2d phalanx in the 5th finger (brachymesophalangia). During growth, the skull presents various typical findings. Among other organ anomalies that occur with increased frequency among mongoloid children are congenital heart disease, frequently as a ventricular septal defect, intestinal atresia, especially in the duodenum, and Hirschsprung's disease.

F has *chondrodystrophia calcificans congenita*, a congenital disturbance of ossification of hitherto unknown etiology, characterized by the presence of punctate sclerotic calcifications in the growing cartilages of the skeleton, mainly in the epiphyses of the long bones, in the primary ossification centers of the hands and feet and in the vertebral column. However, the abnormal mineralization may occur elsewhere; e.g., in the cartilage of the larynx or – as here – of the trachea. The radiologic findings may disappear completely in the course of 2–3 years, but some of the children show disturbances of growth, with permanent and, as a rule, asymmetric shortening of the extremities. Associated malformations have been described, such as dislocation of the hip, clubfoot, flexion contractures, cranial changes, heart disease and bilateral cataract. In severe cases of achondroplasia there may be irregular calcifications in the primary and secondary ossification centers. In the neonate, these may resemble chondrodystrophia calcificans congenita, but these children are easily recognized by the other bone changes characteristic of achondroplasia.

(F-2) is from another patient with chondrodystrophia calcificans congenita.

Recent studies appear to suggest that anticoagulation treatment of the mother during pregnancy may result in bone changes resembling those of chondrodystrophia calcificans congenita.

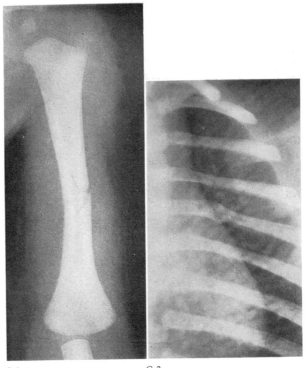

C-2   C-3

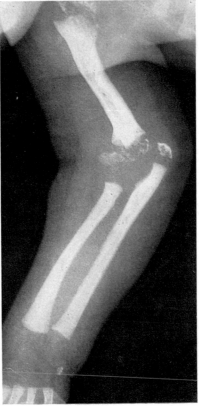

F-2

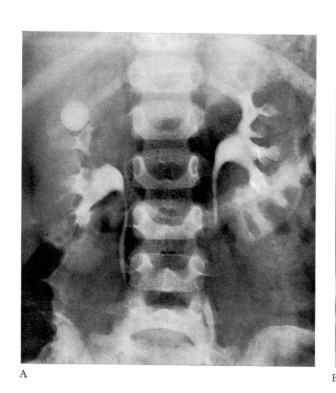

A

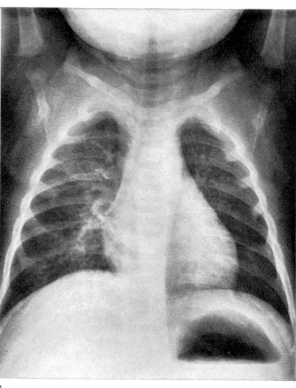

B

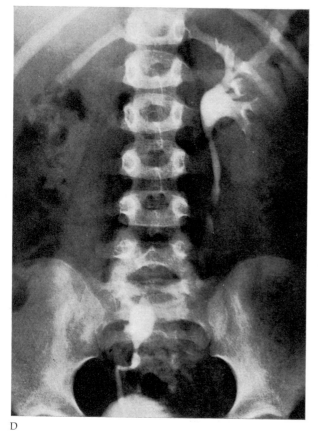

D

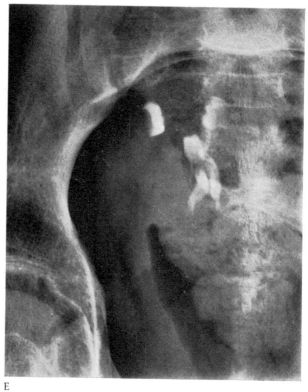

E

Casual Findings

Radiologic studies often reveal lesions of quite another nature than those expected, or lesions localized to organs that were not under clinical suspicion. These may be congenital, nonsymptomatic anomalies or pathologic changes that still have not become manifest clinically or have been overlooked or misinterpreted. Naturally, it also happens that radiologic examination reveals quite another cause of the patient's symptoms than that anticipated.

Here are a few examples of such casual findings. What do the radiograms show? Can the findings be expected to give rise to symptoms at a future date and, if so, why?

A is a 3½-year-old girl, suspected of having a periappendicular abscess. For 6 months she has complained of diffuse intermittent abdominal pain. It disappears after defecation, which is normal. She has never had urinary tract infection, but urine microscopy is performed as part of the investigation, and likewise is found to be normal. Intravenous urography is performed "for safety's sake."

B is a 7-month-old girl, hospitalized for asthmatic bronchitis and a suspicion of pneumonia. During the past 3 days she has had a rise in temperature to 39.5° C, labored respiration and cough. There are distinct perihilar pneumonic consolidations in the right lung, but is she also suffering from something else?

C is a 9-month-old girl, hospitalized for pneumonia. Radiologic examination of the chest reveals bronchopneumonic consolidations in both lungs, but, in addition, it is noticed that the stomach air is lying below the right diaphragmatic leaf. Six days later, when the pneumonia had subsided, radiologic examination of the stomach and the intestine was carried out, followed later by radiologic examination of the colon.

D is a 5-year-old girl, hospitalized for pyuria, anemia and hypersedimentation, demonstrated 3 weeks before admission.

E is a 9-year-old girl, hospitalized for appendicitis. For 6 months she has had at intervals pain in the right lower quadrant of the abdomen, usually lasting for 10-12 hours at a time. On two previous hospitalizations, no explanation of the pain was revealed. All laboratory studies were normal, and after each admission the pain disappeared in the course of a few hours.

F is an 8-year-old girl, hospitalized for recurring infection of the lower urinary tract three times during the past 6 months. On admission, she is free from symptoms and the urine is sterile.

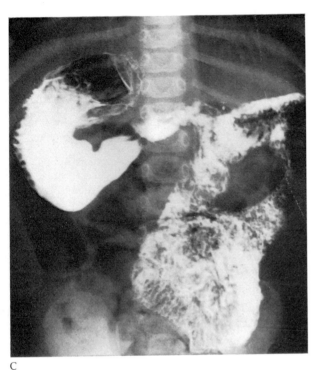

C

F

Answers

A. No explanation was found for the stomach pain, but intravenous urography revealed a round, contrast-filled cavity 12 mm in size, lying immediately laterad to the uppermost calix minor in the right kidney. Renal outlines and pelves are otherwise normal. This is a *caliceal cyst* (caliceal diverticulum, pyelogenic cyst). Caliceal cysts most often are solitary, usually small and communicate with the pelvis. Their etiologic origin probably is different from other types of renal cystic disease. These small cavities probably are without clinical significance, unless the conditions are complicated by infection or the formation of stone.

B has *rickets.* All the bones in the thoracic skeleton and shoulder girdle have slightly ill-defined boundaries and the structure appears blurred because of decalcification. There is only a modest increase in width of the anterior ends of the ribs but more pronounced rachitic changes in the proximal metaphyses of the humerus. Radiologic examination of the wrist *(B-2)* confirmed the diagnosis. The osseous changes disappeared rapidly on treatment with vitamin D.

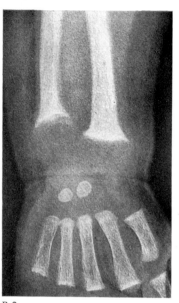

B-2

C has *partial situs inversus.* The stomach therefore lies on the right side, but, in addition, it is seen that the duodenum passes down to the left of the vertebral column and that the loops of jejunum are lying in the left side of the abdomen. There thus is in addition a *rotational anomaly of the intestinal tract,* and subsequent radiologic examination of the colon confirmed that the cecum was lying in front of the vertebral column with most of the large bowel placed in the right side of the abdomen, indicating the presence of a *mesenterium commune.* The heart was placed normally. Partial situs inversus may occur as the sole anomaly, but associated malformations are relatively common, including congenital heart disease. This patient was found to have tachypnea, slight cyanosis, mainly peripheral, as well as a precordial bulge but no electrocardiographic changes. There was uncharacteristic enlargement of the heart and increased pulmonary vascularity, suggesting the presence of a left-to-right shunt.

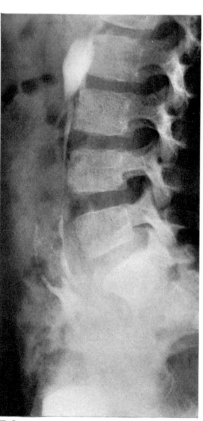

D-2

D does not lack the *right kidney,* but it is *ectopic.* It has not ascended normally and is lying in front of the sacral promontory *(D-2).* Neither this nor other anomalies in the position of the kidneys, such as crossed ectopia, malrotation or horseshoe kidney, have any significance provided that the development of the kidney is normal and the flow is free. However, the anomalies often cause an obstruction to flow, with development of hydronephrosis and a disposition to infection and reflux. In this patient, the ectopic kidney appears well developed, with free flow through the ureter, but in the course of a possible future pregnancy there is a grave risk of obstruction by pinching the ureter.

E, in the course of her third hospitalization, underwent radiologic examination of the stomach and colon, which were normal. However, in the right side of the pelvis some small calcifications were noticed, so a plain film of the pelvic region was made. The radiogram shows a number of intense, homogeneous and well-delimited calcifications with the appearance of teeth. *E* has a *dermoid cyst* and at operation a pedunculated tumor was removed, almost as large as a tennis ball, arising from the right ovary. It was twisted several times on its own stalk, without the vascularization being apparently compromised. It is probable that the twisting has been the cause of the intermittent abdominal pain.

On cutting the preparation, a thick-walled cyst was found containing large amounts of sebaceous material, hair and cartilage with several teeth.

F has a *horseshoe kidney, ren arcuatus,* with fusion of the inferior poles. As is characteristic for this anomaly, the renal pelves turn forward just as in nonrotated kidneys. No obstruction to flow can be demonstrated in this patient, so that the anomaly has hardly any significance for the occurrence of urinary tract infection. However, pinching of one or both ureters often is seen during their passage down across the fused lower poles, with the development of hydronephrosis. *(F-2)* is such a case in a 7-year-old girl with pyuria and recurrent pain in the loins.

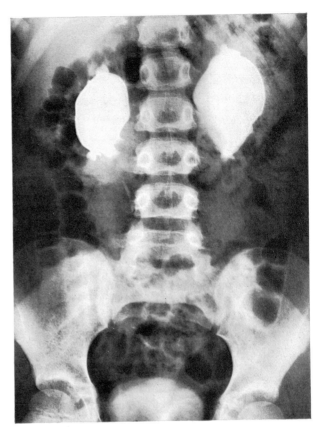

F-2

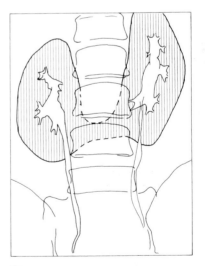

(F)

Index

(with references to the film – page and designation of film)